ENVIRONMENTAL SCIENCE, ENGINEERING AND TECHNOLOGY

ECOSYSTEM MODELING AND ITS APPLICATION FOR SEAGRASS BEDS

ENVIRONMENTAL SCIENCE, ENGINEERING AND TECHNOLOGY

Additional books in this series can be found on Nova's website under the Series tab.

Environmental Science, Engineering and Technology

Ecosystem Modeling and its Application for Seagrass Beds

Akio Sohma

Nova Science Publishers, Inc.
New York

Copyright ©2011 by Nova Science Publishers, Inc.

All rights reserved. No part of this book may be reproduced, stored in a retrieval system or transmitted in any form or by any means: electronic, electrostatic, magnetic, tape, mechanical photocopying, recording or otherwise without the written permission of the Publisher.

For permission to use material from this book please contact us:
Telephone 631-231-7269; Fax 631-231-8175
Web Site: http://www.novapublishers.com

NOTICE TO THE READER

The Publisher has taken reasonable care in the preparation of this book, but makes no expressed or implied warranty of any kind and assumes no responsibility for any errors or omissions. No liability is assumed for incidental or consequential damages in connection with or arising out of information contained in this book. The Publisher shall not be liable for any special, consequential, or exemplary damages resulting, in whole or in part, from the readers' use of, or reliance upon, this material. Any parts of this book based on government reports are so indicated and copyright is claimed for those parts to the extent applicable to compilations of such works.

Independent verification should be sought for any data, advice or recommendations contained in this book. In addition, no responsibility is assumed by the publisher for any injury and/or damage to persons or property arising from any methods, products, instructions, ideas or otherwise contained in this publication.

This publication is designed to provide accurate and authoritative information with regard to the subject matter covered herein. It is sold with the clear understanding that the Publisher is not engaged in rendering legal or any other professional services. If legal or any other expert assistance is required, the services of a competent person should be sought. FROM A DECLARATION OF PARTICIPANTS JOINTLY ADOPTED BY A COMMITTEE OF THE AMERICAN BAR ASSOCIATION AND A COMMITTEE OF PUBLISHERS.

Additional color graphics may be available in the e-book version of this book.

LIBRARY OF CONGRESS CATALOGING-IN-PUBLICATION DATA

Sohma, Akio.
  Ecosystem modeling and its application for seagrass beds / Akio Sohma.
    p. cm.
  Includes index.
  ISBN 978-1-61209-290-4 (softcover)
  1. Seagrasses--Habitat. 2. Habitat (Ecology)--Japan--Mikawa Bay. I. Title.
  SH393.S64 2011
  584'.74--dc22
                              2010054287

Published by Nova Science Publishers, Inc. † New York

Contents

Preface		vii
Chapter 1	Introduction	1
Chapter 2	The Environmental Assessment Project of Atsumi Bay	5
Chapter 3	Ecosystem Modeling	11
Chapter 4	Implementation	39
Chapter 5	Evaluation of Part 1 - Likelihood of Seagrass Growth on the Artificial Shallows	53
Chapter 6	Evaluation of Part 2 - Effect of Reclaiming/ Creating the Seagrass Beds on Atsumi Bay	73
Chapter 7	Evaluation of the Purification	85
Chapter 8	Conclusion	93
Acknowledgment		95
References		97
Index		107

PREFACE

An ecosystem model was developed and applied to Atsumi Bay, Japan, to evaluate the effects of reclaiming seagrass beds and creating artificial shallows with seagrass beds to mitigate the effects of the reclamation. The model can demonstrate the ecological mechanisms in the pelagic and benthic ecosystems including seagrass beds and tidal flats. The objectives of the application of the ecosystem model to Atsumi Bay are (a) to investigate the likelihood of cultivating and maintaining seagrass beds in artificial shallows (Part 1), and (b) to understand the effects of the reclamation of seagrass beds and the creation of artificial shallows on the water quality in the estuary (Part 2). In Part 1, first, the nutrient turnover rates due to both biochemical and physical processes in the natural seagrass beds where reclamation is proposed were analyzed. Biological processes rather than physical processes were the most significant driving forces of nutrient cycles in seagrass beds. Second, the effects of filter feeding benthic fauna (suspension feeders) in the seagrass beds were analyzed. The scenario with suspension feeders resulted in higher transparency of the water column (8.7% decrease in the light attenuation coefficient) and an increase in nutrient supply (24.9% increase in NH_4-N in the water column) contributing to the high specific growth rate of seagrass. Third, the specific growth rate of seagrass on the proposed artificial shallows was measured. The value on the artificial shallows set at a depth of Datum Line minus 0.8 m (D.L. -0.8 m) was approximately the same as that of the natural seagrass beds. In Part 2, first, water quality in the estuary was compared among the scenarios with/without natural seagrass beds and artificial shallows. Then, the defined values of the water purification capability of (a) artificial shallows with/without seagrass beds, and (b) natural seagrass beds per unit area were evaluated. The reclamation of the natural seagrass beds resulted in a decrease of the removal of phytoplankton and detritus from the pelagic system (i.e. resulted in a loss in the purification rate). In contrast, the creation of artificial shallows resulted in an increase of the removal of phytoplankton and detritus from the pelagic system (i.e. resulted in a gain in the purification rate). Based on an annual average, approximately twice as much phytoplankton was removed from the artificial shallows at the depth D.L. -0.8 m, than at the depth, D.L. -1.5 m, and the artificial shallows with seagrass beds removed pelagic DIN and DIP at a rate 120% higher than that without seagrass beds.

Chapter 1

INTRODUCTION

Seagrass bed ecosystems usually have the most complicated aquatic cultures in estuaries due to living the many speices and quantity of biologies and is the highest biological production areas in the estuary. The functions of seagrass beds are known as (1) providing nursery ground for marine biologies, (2) driving nutrients up from lower trophic to higher trophic level through food web smoothly and (3) preventing red tides (rapid growth of phytoplankton) and hypoxia through the ecological/physical interaction among benthic-pelagic ecosystems and central bay-shallow water areas. The research introduced here is focused on item (3).

Usually, the location of seagrass beds is at a part of shallow waters where enough light achieves to the water column and sea floor for its photosynthesis. In shallow waters, biodeposits from filter feeding benthic fauna (suspension feeders) contribute to the total suspended load in eutrophic coastal zones (Haven and Morales-Alamo, 1966; Tenore and Dunstan, 1973; Kraeuter, 1976; Tsuchiya, 1980). Such biologically mediated sedimentation has the capacity to exceed passive physical processes in the deposition of fresh sediments in estuaries. In addition, these biodeposits represent a potentially significant energy source to consumers such as seagrass beds (Peterson and Heck Jr., 1999). The mineralization of organic matter by benthic fauna supplies the nutrients for the photosynthesis of seagrasses. These ecological mechanisms potentially transfer planktonic production from the water column to the benthos through feces and pseudo feces, and thereby enhance submerged aquatic vegetation growth by increasing the nutrients available in the rhizosphere. This transfer of production in shallow waters prevents the rapid growth of phytoplankton (red tides) directly and also prevents the increase of detritus sedimentation both in offshore areas and in central bay areas where hypoxia has been frequently observed.

In spite of the demonstrated importance of shallows in preventing the coastal environmental problems related to eutrophication, shallows, especially those close to urban areas, are also economically attractive sites for reclamation for industrial activity. Given these conflicting benefits and uses of shallow waters, the strategic plans for operation and management are urgently required in Japan. To implement suitable operation and management practices, it is necessary (1) to understand the ecological mechanisms of seagrass beds and (2) to predict/estimate the effects of environmental/development measures.

The ecosystem model is a powerful tool to evaluate complex systems like the shallows/seagrass beds which are influenced by many biochemical and physical processes. It can be used to develop our understanding of the internal ecosystem mechanisms and for predicting the ecosystem response to environmental measures. As a result, the model is likely a useful tool not only for scientific purposes, but also for a communication platform of operation and management of coastal zones (Sohma et al., 2009, Sohma, 2010).

Coastal ecosystem modeling began as the modeling of the pelagic ecosystem (e.g. Kremer and Nixon, 1987). The model was then expanded to the benthic ecosystem (Boudreau, 1996, Soetaert et al., 1996, Sohma and Sayama, 2002), tidal flats ecosystem (Baretta and Ruardij, 1988, Sohma et al., 2000) and the coupling of the benthic-pelagic ecosystems (Baretta et al., 1995, Baretta-Bekker and Baretta, 1997, Sohma et al., 2001, Luff and Moll, 2004). Concerning seagrass beds, Hata and Nakata (1998) developed an ecosystem model focused on the nitrogen cycle of eelgrass beds. However, these mentioned models did not focus on the interaction between the areas of seagrass beds and central bay areas which often become oxygen depleted.

More recently, a benthic-pelagic coupled ecosystem model named "TRÄUMEREI" was developed which represents the coastal area comprised of seagrass beds, shallow waters/tidal flats without seagrass, and oxygen depleted areas. The model was applied to the Jinno area including the proposed reclamation site. In the study (Sohma et al., 2004), the model was validated and could reproduce the seasonal and daily variations of the ecosystem in the Jinno area, and the model illustrated the important role of benthic-pelagic interactions through the benthic fauna and seagrass beds (biodeposition and mineralization of organisms by suspension feeders and the utilization of nutrients by vegetation). In addition, the New Coastal Marine Ecosystem Model (the NCME model), an earlier version of the "TRÄUMEREI" model, was developed, which modeled the benthic and pelagic ecosystems of the estuary where light transparency is so low that it hinders the growth of seagrasses, seaweeds, and benthic algae. The NCME model was applied to and validated for Mikawa Bay, which includes Atsumi Bay (Sohma et al., 2001). In that earlier study, the NCME model reproduced the seasonal and daily variations of the ecosystem in Mikawa Bay.

The objective of this document is to introduce (A) the ecosystem model, "TRÄUMEREI" and (B) the evaluation programs for the reclamation and mitigation plans for seagrass beds by using the ecosystem models "TRÄUMEREI" and "NCME". Item (A) and item (B) were developed in "the environmental assessment project of Atsumi Bay", Japan, and these developments were executed simultaneously because of their strong relationship.

Chapter 2

THE ENVIRONMENTAL ASSESSMENT PROJECT OF ATSUMI BAY

2.1. LOCATION AND SITUATION

Atsumi Bay (34°70'N, 137°20'E) is one of the most eutrophic and hypoxic estuaries in Japan. There is a reclamation plan for the Jinno area, where seagrass beds (eelgrass, *Zostera marina*) are present (Figure 1). To alleviate the environmental impacts of the reclamation of these seagrass beds in the Jinno area, several mitigation plans are proposed. One plan is to create artificial shallows in the Mito area (Figure 1), where an oxygen depleted water body is frequently observed. The expectations for the artificial shallows in the Mito area are: that they will promote the natural growth of seagrass beds and associated benthic fauna, and they will restore the ecological function lost due to the reclamation of the seasgrass beds in the Jinno area. Quantitative evaluation is urgently required for the effect of the reclamation plan and the performance of mitigation plans.

2.2. EVALUATION PROGRAM

The developed evaluation program for the reclamation plan of seagrass beds in the Jinno area and the mitigation plan for the creation of artificial shallows in the Mito area was divided into two major parts, "Part 1" and "Part 2". An overview of the application of the ecosystem model for both parts is given below.

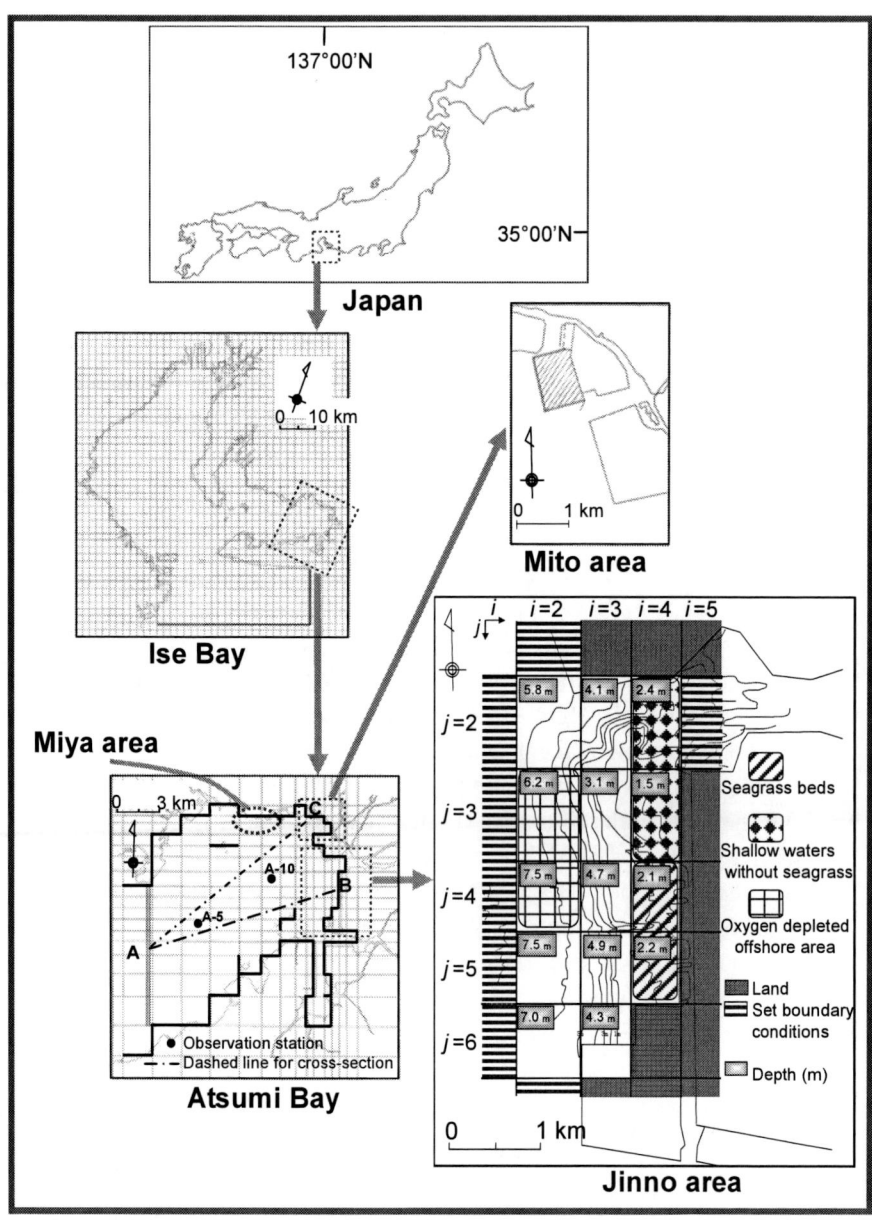

Figure 1. Location of the study area (Atsumi Bay, Jinno area, Mito area, and Miya area) -natural seagrass beds are situated in the Jinno area, (i,j)=(4,4) and (4,5).

2.2.1. Objective and Scenarios for Part 1

The objective of Part 1 is the evaluation of the likelihood of seagrass growth on the new artificial shallows. In Part 1, the ecosystem model, TRÄUMEREI (Sohma et al., 2004) described details later was applied in three scenarios (Table 1). The first scenario, 1-1, is the existing condition in the Jinno area comprised of not only seagrass beds but also shallows/tidal flats without seagrass and oxygen depleted areas as shown in Figure 1. The aims of scenario 1-1 are (a) to validate the TRÄUMEREI model, (b) to understand the characteristics of the biological mechanisms in the natural seagrass beds through the comparison of three different areas (i.e. seagrass beds, shallow/tidal flats without seagrass and oxygen depleted areas), and (c) to assess which factors are important for the growth of seagrass beds. The second scenario, 1-2, is where we assess the absence of suspension feeders in the Jinno area. The aim of scenario 1-2 is to assess the effect of the biological processes attributed to suspension feeders (feeding, feces, excretion etc.) on the growth rate of seagrass (specific growth rate due to photosynthesis). Scenario 1-2 was designed based on the results of scenario 1-1, which highlighted the high contribution rate of the biological activity of suspension feeders to the nutrient cycle in seagrass beds (discussed in section 5.1). The third scenario, 1-3, is the creation of artificial shallows with seagrass beds in the Mito area. The seagrass growth rate in scenario 1-3 is compared to that of scenario 1-1. The aim of scenario 1-3 is to quantitatively evaluate the growth of seagrass beds on artificial shallows in the Mito area.

Table 1. Scenarios in Part 1 - for the evaluation of the likelihood of seagrass growth on the new artificial shallows

Applied Area	Scenario	Scenario Objectives
Jinno area	• Scenario 1-1: Existing Condition	• Validation of the TRÄUMEREI model. • Evaluation of the characteristics of the mechanism in natural seagrass beds. • Clarification of important factors to seagrass growth.
Jinno area	• Scenario 1-2: Assumption: Absence of suspension feeders	• Evaluation of the effectiveness of the biological processes of suspension feeders to seagrass growth.
Mito area	• Scenario 1-3: Assumption: Creation of artificial shallows with seagrass	• Evaluation of the likelihood of seagrass growth on the artificial shallows at Mito area.

Table 2. Scenarios in Part 2 - for the evaluation of the effect of reclaiming seagrass beds in the Jinno area and creating artificial shallow waters in the Mito area on the water quality of the estuary, Atsumi Bay

Scenario	Settings for the Jinno area	Settings for the Mito area	Scenario Objectives
Scenario 2-1	• *Ecological condition*: Existence of seagrass beds, epiphytes and epifauna (Existing condition) • *Geographical condition*: No reclamation (Existing condition)	• *Ecological condition*: Hypoxic ecosystem (Existing condition) • *Geographical condition*: No creation of artificial shallows (Existing condition)	• Validation of the model coupling TRAUMEREI and NCME models. • Clarification of the differences between existing condition and the four scenarios described below.
Scenario 2-2	• *Ecological condition*: No seagrass, epiphytes and epifauna • *Geographical condition*: No reclamation (Existing condition)	• *Ecological condition*: Hypoxic ecosystem (Existing condition) • *Geographical condition*: No creation of artificial shallows (Existing condition)	• Clarification of the effect of seagrass beds in the Jinno area on Atsumi Bay.
Scenario 2-3	• *Ecological condition*: No seagrass, epiphytes and epifauna • *Geographical condition*: Reclamation of seagrass beds (Reclaimed condition)	• *Ecological condition*: Ecosystem of shallows without seagrass • *Geographical condition*: Creation of artificial shallows (Condition of mitigation plan)	• Evaluation of the effect of mitigation plan in the Mito area towards the reclamation in the Jinno area. (The failed case of seagrass cultivation on the artificial shallows in Mito area).
Scenario 2-4	• *Ecological condition*: No seagrass, epiphytes and epifauna • *Geographical condition*: Reclamation of seagrass beds (Reclaimed condition)	• *Ecological condition*: Ecosystem of shallows with seagrass • *Geographical condition*: Creation of artificial shallows (Condition of mitigation plan)	• Evaluation of the effect of mitigation plan in the Mito area towards the reclamation in the Jinno area. (The successful case of seagrass cultivation on the artificial shallows in the Mito area.)
Scenario 2-5	• *Ecological condition*: No seagrass, epiphytes and epifauna • *Geographical condition*: Reclamation of seagrass beds (Reclaimed condition)	• *Ecological condition*: Hypoxic ecosystem (Existing condition) • *Geographical condition*: No creation of artificial shallows (Existing condition)	• Clarification of the effect of the mitigation plan in the Mito area towards the reclamation in the Jinno area. (The comparison between scenario 2-3 and 2-5, or between scenario 2-4 and 2-5.)

2.2.2. Objective and Scenarios for Part 2

The objective of Part 2 is the evaluation of the effects of reclaiming the seagrass beds in the Jinno area and creating artificial shallows in the Mito area on the water quality in Atsumi Bay. In Part 2, the ecosystem model coupling the TRÄUMEREI and the NCME models (Sohma et al., 2001, Sohma et al., 2004) was applied to the five scenarios shown in Table 2. The TRÄUMEREI model was applied to the Jinno area and the NCME model was applied to Atsumi Bay excluding the Jinno area. The first calculated scenario, 2-1, represents the existing conditions in Atsumi Bay (Figure 1). The aims of scenario 2-1 are (a) to validate the coupling of the TRÄUMEREI and the NCME models and (b) to clarify the differences between the existing conditions and the four scenarios described hereafter. The second scenario calculated, 2-2, is for Atsumi Bay assuming no seagrass beds in the Jinno area. The aim of scenario 2-2 is to investigate the impact of the seagrass beds in the Jinno area on Atsumi Bay by comparing two model runs, i.e. scenario 2-2 and scenario 2-1. The third scenario calculated, 2-3, is for Atsumi Bay assuming both the creation of artificial shallows in the Mito area and the reclamation of seagrass beds in the Jinno area. In this scenario, the seagrass bed on the artificial shallows in the Mito area is set not to shoot (i.e. not to grow) and the biomass of seagrass is set at zero. The fourth scenario calculated, 2-4, is for Atsumi Bay with the creation of artificial shallows with the growth of seagrass beds in the Mito area and with the reclamation of the Jinno area (as for scenario 2-3). The fifth scenario calculated, 2-5, is for Atsumi Bay assuming the existing conditions in the Mito area (no mitigation plan) and with the reclamation in the Jinno area. Comparisons between scenarios 2-3 and 2-5, or between scenarios 2-4 and 2-5 demonstrate the effect of the mitigation plan in the Mito area (the creation of artificial shallows without or with seagrass beds) on the reclamation in the Jinno area. The comparison between scenarios 2-4 and 2-5 is the evaluation for the successful case of seagrass growth on the artificial shallows at Mito area, while the comparison between application 2-3 and 2-5 is the evaluation for the failed case of seagrass cultivation on it.

Chapter 3

ECOSYSTEM MODELING

3.1. WHOLE CONSTRUCTION OF THE SIMULATION SYSTEM

The simulation system is composed of two models, the ecosystem model and the hydrodynamics model.

The hydrodynamic model simulates the three dimensional physical field in the estuary and demonstrates the long-term variability of flow field, and heat transport. The numerical development and algorithm of the hydrodynamic model are well described by Nakata et al (1983) and Sohma (2010). The model includes the tidal forcing, surface wind and local density gradient together with the realistic coastal topography and bathymetry. Model equations are based on fluid motion, flow continuity and conservation of heat and salt. The vertical mixing process is parameterized with a turbulence model of second moment closure, which determines local distributions of the turbulent kinetic energy, k, and the mixing length, l, by means of well-established k-kl equations (e.g., Blumberg and Mellor, 1978; Mellor and Yamada, 1982).When the ecosystem model is calculated, outputs of the hydrodynamics model at each time-step (flow velocity, temperature, turbulent kinetic energy, tidal level, etc.) are used. By coupling ecosystem model with the outputs of the hydrodynamic model, this simulation system can estimate the water quality of the estuary including the effect of physical processes completely.

Concerning with ecosystem model, the system has two ecosystem models, the TRÄUMEREI model and the NCME model. As described in "*1. Introduction*", the TRÄUMEREI model is improved version of the NCME model. The NCME model targets the estuary where transparency is so low that it hinders sea-grass, seaweed, and benthic algae photosynthesis. As a result, the model did not include the following model variables, i.e. seagrass, seaweed, and benthic algae, plus their biological processes. The new version of the ecosystem model, TRÄUMEREI, adds following variables and processes to the NCME model. Those are seagrass, seaweed, benthic algae, epihpytes, epifauna, and their biochemical processes plus the effects of emersion/submersion of tidal flats.

The TRÄUMEREI model includes all variables/processes treated in the NCME model, and the model concept, construction and equation are basically the same as NCME model. The details of TRÄUMEREI are introduced in the next section.

3.2. Ecosystem Model -TRÄUMEREI

3.2.1. Overview of the TRÄUMEREI Model

The TRÄUMEREI model was developed to evaluate both the physical and biochemical processes in the estuarine lower trophic ecosystem in terms of coupling of carbon, nitrogen, phosphorus and oxygen cycles. The major material fluxes that are treated in each mesh/box of the model are shown in Figure 2. Various forms that these elements take are treated as variables named 'model components' shown as orange boxs in Figure 2 and the temporal and spatial dynamics of model components are described by partial differential equations based on mass conservation of carbon, nitrogen, phosphorus and oxygen elements. In these equations, physical and biochemical processes of both the pelagic and the benthic system are included. As for the physical processes, pelagic substances in the model are influenced by advection from water transport and diffusion (eddy viscosity) from turbulent flow. Particulate substances of benthic systems in the model are influenced by advection (sediment deposition) and activity of organisms (particle mixing; bioturbation). Bioturbation is modeled as a diffusion-like process. Dissolved substances of benthic system in the model move by molecular diffusion, bioturbation and irrigation by benthic macrofauna and advection from pore water velocity. Flow velocity and eddy viscosity data of seawater in the equations is derived from the calculation of the hydrodynamic model and are implemented to the TRÄUMEREI model. As for the biochemical processes, details of treated processes in the model are described in the section of "*3.2.3. Biochemical Reactions*".

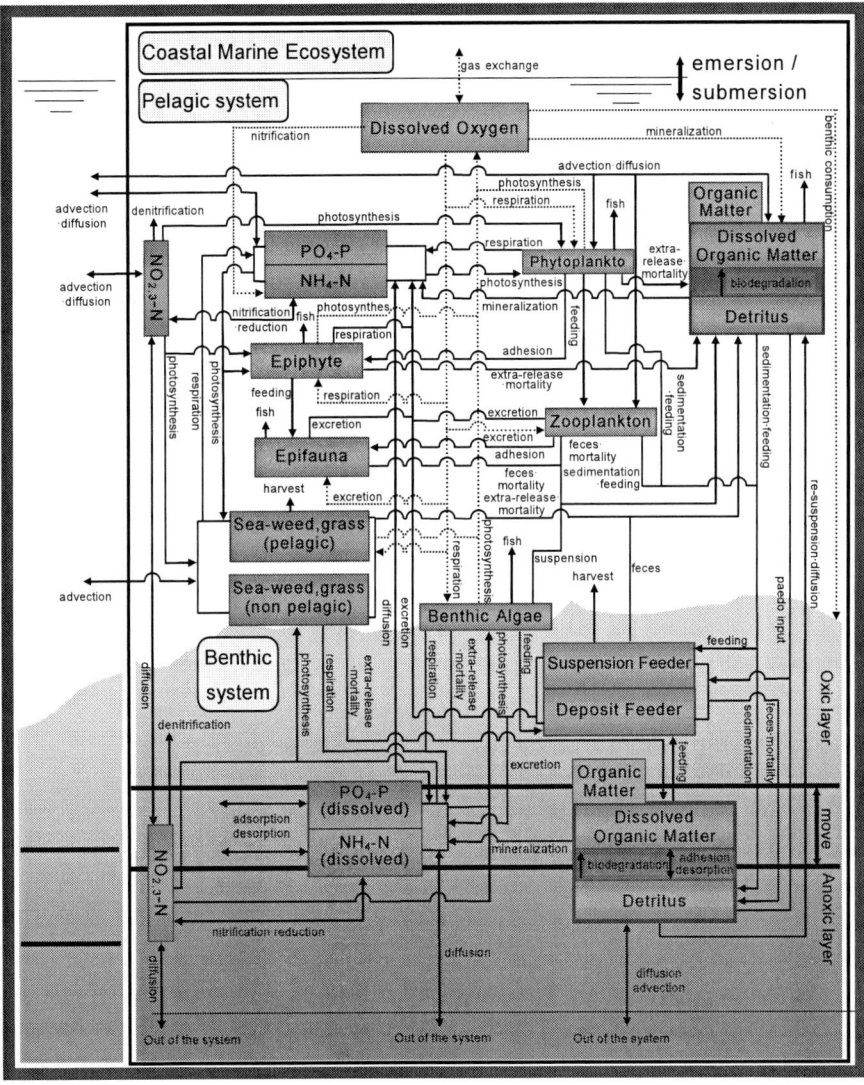

Figure 2. Coupled cycle of carbon, nitrogen, phosphorous, and oxygen, in one mesh of new ecosystem model (the TRÄUMEREI). Model components are represented with box.

The temporal and spatial dependency of model components are decided at each time step as a result of the interactions (= physical and biochemical processes) among model components or among model components and environmental variables. Here, environmental variable is defined as the one, which becomes the driving force of the cycles of matter (carbon, nitrogen, phosphorus, and oxygen) and is described by a prescribed function; for example, intensity of light. The new version of the ecosystem model (TRÄUMEREI) was developed based on the NCME model. Additional factors of the TRÄUMEREI model from the NCME model (Sohma et al 2001) are several model components; benthic algae, sea-grass, seaweed, epiphytes and epifauna and biochemical processes associated with these components. Furthermore, the effects of tidal flat emersion/submersion on biochemical processes are also included. The period of the emerged time of each mesh/box is calculated from the tidal level that was pre-calculated in the hydrodynamics model. In the emerged mesh, the model can take into account several differences affecting the ecological mechanism viz: (1) boundary flux at the sediment-water interface; (2) feeding by suspension feeders; and (3) high temperature of sediment compared to water especially during the daytime in summer. When we actually ran the model, while the mesh are emerged, the fluxes across the sediment-water interface such as diffusion of dissolved substances, sediment deposition and feeding of suspension feeder were set at zero and boundary conditions of oxygen at sediment interface were set as saturated concentration.

3.2.2. Model Equations and Computational Algorithm

The model equations and computational schemes based are presented here.

(1) For Pelagic System
The general equations for pelagic system is as follows:

$$\frac{\partial C_w}{\partial t} = -(\mathbf{v} \bullet \nabla)C_w + \nabla \bullet (\mathbf{K} \bullet \nabla C_w) + \sum R$$

$$= -u\frac{\partial C_w}{\partial x} - v\frac{\partial C_w}{\partial y} - w\frac{\partial C_w}{\partial z}$$

$$+ \frac{\partial}{\partial x}\left(K_x \frac{\partial C_w}{\partial x}\right) + \frac{\partial}{\partial y}\left(K_y \frac{\partial C_w}{\partial y}\right) + \frac{\partial}{\partial z}\left(K_z \frac{\partial C_w}{\partial z}\right) + \sum R \quad (3.2.1)$$

where, C_w = model components (mass/L^3-liquid), $\mathbf{v} = (u, v, w)$ = flow velocity that already has been calculated by the hydrodynamics model (L/T), t

= time (T), x, y, z = spatial coordinates (L), $\sum R$ = biochemical reactions and fluxes (mass/L^2-liquid/T), $K = \begin{pmatrix} K_x & 0 & 0 \\ 0 & K_y & 0 \\ 0 & 0 & K_z \end{pmatrix}$ = eddy diffusion (viscosity) tensor (L^2-liquid/T).

Assuming *div* **v** = 0, we obtained model equations for pelagic system.

$$\frac{\partial C_w}{\partial t} = -\frac{\partial(u C_w)}{\partial x} - \frac{\partial(v C_w)}{\partial y} - \frac{\partial(w C_w)}{\partial z}$$
$$+ \frac{\partial}{\partial x}\left(K_x \frac{\partial C_w}{\partial x}\right) + \frac{\partial}{\partial y}\left(K_y \frac{\partial C_w}{\partial y}\right) + \frac{\partial}{\partial z}\left(K_z \frac{\partial C_w}{\partial z}\right) + \sum R \quad (3.2.2)$$

Temporal-spatial distribution of Pelagic model components, phytoplankton, zooplankton, detritus (fast labile POM, slow labile POM, refractory POM), dissolved organic matter (labile DOM, refractory DOM), NH_4-N, $NO_{2,3}$-N, PO_4-P, DO, and ODU, are calculated by equation 3.2.2. From the spatial integration of equation 3.2.2, the model equation is obtained to use finite volume method (FVM), whose algorithm conserved mass of components. The algorithm and the coordination are shown in Table 3 and Figure 3.

Table 3. Model equation and algorithm in the pelagic ecosystem (refer to Figure 3)

Surface layer in the pelagic system=free surface: (i,j,1) Mesh/Box
From the spatial integration of (i,j,1) mesh/box

$$\iiint_{\varsigma(t)}^{z(2)} \frac{\partial}{\partial t} C \, dxdydz + \iiint_{\varsigma(t)}^{z(2)} \frac{\partial}{\partial x}\left(uC - K_x \frac{\partial C}{\partial x}\right) dxdydz$$
$$+ \iiint_{\varsigma(t)}^{z(2)} \frac{\partial}{\partial y}\left(vC - K_y \frac{\partial C}{\partial y}\right) dxdydz + \iiint_{\varsigma(t)}^{z(2)} \frac{\partial}{\partial z}\left(wC - K_z \frac{\partial C}{\partial z}\right) dxdydz$$
$$= \iiint_{\varsigma(t)}^{z(2)} \sum R \, dxdydz \quad (T3\text{-}1)$$

$\Leftrightarrow$

Table 3. (Continued).

$$\frac{\partial}{\partial t}\left(\Delta x \cdot \Delta y \cdot \int_{\zeta(t)}^{z(2)} C_P \cdot dz\right) + \Delta x \cdot \Delta y \cdot \frac{\partial \zeta}{\partial t} C\bigg|_{z=\zeta(t)} + \Delta y \cdot \left(\int_{\zeta(t)}^{z(2)} F_e \, dz - \int_{\zeta(t)}^{z(2)} F_w \, dz\right) + \Delta x \cdot \left(\int_{\zeta(t)}^{z(2)} F_s \, dz - \int_{\zeta(t)}^{z(2)} F_n \, dz\right)$$

$$+ \Delta x \cdot \Delta y \cdot \left(F_{dn} - F_{up}\bigg|_{z=\zeta(t)}\right) = \Delta x \cdot \Delta y \cdot \Delta z \cdot \sum R_P \quad \text{(T3-2)}$$

$$\Leftrightarrow$$

$$\frac{\Delta x \cdot \Delta y \cdot (\Delta z - \zeta(t+\Delta t)) \cdot C_P^{t+\Delta t} - \Delta x \cdot \Delta y \cdot (\Delta z - \zeta(t)) \cdot C_P^t}{\Delta t} + \Delta y \cdot \left\{(\Delta z - \zeta(t)) \cdot F_e^t - (\Delta z - \zeta(t)) \cdot F_w^t\right\}$$

$$+ \Delta x \cdot \left\{(\Delta z - \zeta(t)) \cdot F_s^t - (\Delta z - \zeta(t)) \cdot F_n^t\right\} - \Delta x \cdot \Delta y \cdot F_{dn}^{t+\Delta t} = \Delta x \cdot \Delta y \cdot \Delta z \cdot \sum R_P^t \quad \text{(T3-3)}$$

Any layer without surface in the pelagic system: (i,j,k) Mesh/Box
from the spatial integration of (i,j,k) mesh/box

$$\iiint \frac{\partial}{\partial t} C \, dxdydz + \iiint \frac{\partial}{\partial x}\left(uC - K_x \frac{\partial C}{\partial x}\right) dxdydz$$

$$+ \iiint \frac{\partial}{\partial y}\left(vC - K_y \frac{\partial C}{\partial y}\right) dxdydz + \iiint \frac{\partial}{\partial z}\left(wC - K_z \frac{\partial C}{\partial z}\right) dxdydz = \iiint \sum R \, dxdydz \quad \text{(T3-4)}$$

$$\Leftrightarrow$$

$$\frac{\partial}{\partial t}\left(\Delta x \cdot \Delta y \cdot \int_{z(k)}^{z(k+1)} C_P \cdot dz\right) + \Delta y \cdot \left(\int_{z(k)}^{z(k+1)} F_e \, dz - \int_{z(k)}^{z(k+1)} F_w \, dz\right) + \Delta x \cdot \left(\int_{z(k)}^{z(k+1)} F_s \, dz - \int_{z(k)}^{z(k+1)} F_n \, dz\right)$$

$$+ \Delta x \cdot \Delta y \cdot (F_{dn} - F_{up}) = \Delta x \cdot \Delta y \cdot \Delta z \cdot \sum R_P \quad \text{(T3-5)}$$

$$\Leftrightarrow$$

$$\frac{\Delta x \cdot \Delta y \cdot \Delta z \cdot C_P^{t+\Delta t} - \Delta x \cdot \Delta y \cdot \Delta z \cdot C_P^t}{\Delta t} + \Delta y \cdot \Delta z \cdot (F_e^t - F_w^t)$$

$$+ \Delta x \cdot \Delta z \cdot (F_s^t - F_n^t) - \Delta x \cdot \Delta y \cdot (F_{dn}^{t+\Delta t} - F_{up}^{t+\Delta t}) = \Delta x \cdot \Delta y \cdot \Delta z \cdot \sum R_P^t \quad \text{(T3-6)}$$

where:

$$F_s = \left(vC - k_y \frac{\partial C}{\partial y}\right)_{y=s}, \quad F_e = \left(uC - k_x \frac{\partial C}{\partial x}\right)_{x=e}, \quad F_{up} = \left(wC - k_z \frac{\partial C}{\partial z}\right)_{z=up}, \quad \ldots$$

$$\zeta(t) \le z(2) \quad \text{for } \forall t$$

Where the model used the implicit scheme in the vertical direction, and applied the high-lateral flux modification (Spalding, 1972; Patankar, 1980) both in the vertical and horizontal fluxes. The model solves this equation by Tri-Diagonal Matrix Algorithm (TDMA).

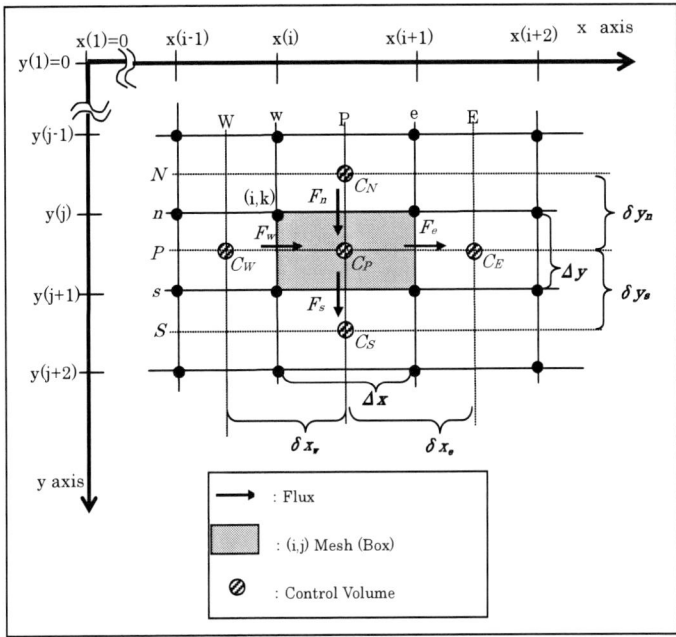

In the horizontal direction

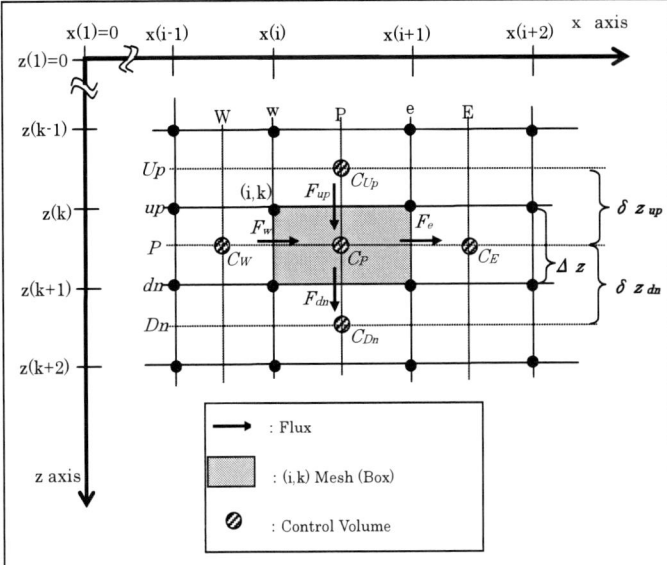

In the vertical direction

Figure 3. Model coordinates in the horizontal direction (upper) and the vertical direction (lower) in the ecosystem model.

(2) For Benthic System

The equations and the computational scheme for benthic system in the TRÄUMEREI model have several improved factors from the NCME model in order to include the effects of differences of porosity in vertical directions and to clear the contribution rate of several diffusion processes from different causes. To lead the model equations for benthic system, the following two diagenetic equations (equations 3.2.3 and 3.2.4) are used. Equations 3.2.3 and 3.2.4 add "two terms" to equations figured by Berner (1980) for the equations to be able to use more generality/variedly. The terms are interphase mixing of biodiffusion (bioturbation), $\partial\{\phi D'_B \partial C/\partial z\}/\partial z$, and irrigation coefficient, $\phi\alpha(C_0 - C)$. Formulations of these two processes are expressed by Boudreau (1986). As the result of including these formulations, there are two selectable expressions about each effect of bioturbation and irrigation processes in our model.

(a) Diagenetic equation for dissolved material

$$\frac{\partial(\phi C)}{\partial t} = \frac{\partial\left\{D_B \frac{\partial(\phi C)}{\partial z} + \phi(D_S + D_I + D'_B)\frac{\partial C}{\partial z}\right\}}{\partial z} + \phi\alpha(C_0 - C)$$

$$-\frac{\partial(\phi v C)}{\partial z} + \phi R_{ads} + \phi \sum R' \qquad (3.2.3)$$

where, ϕ = porosity (-), C = concentration of dissolved substances (mass/L^3-liquid), C_0 = concentration of components at sediment-water interface (mass/L^3-liquid), D_S = molecular diffusion coefficient in sediment including the effects of tortuosity (L^2-sediment/T), D_B = solid biodiffusion coefficient (intraphase mixing expression) (L^2-sediment/T), D'_B = solid biodiffusion coefficient (interphase mixing expression) (L^2-sediment/T), D_I = irrigation coefficient (diffusion-like expression) (L^2-sediment/T), α = irrigation coefficient2 (1/T), v = velocity of burial of water below the sediment-water interface (L-sediment/T), R_{ads} = reactions of dissolved materials due to equilibrium adsorption or desorption (mass/L²-liquid/T). $\sum R'$ = all other slow (irreversible) biochemical reactions (mass/L²-liquid/T).

(b) Diagenetic equation for particulate material

$$\frac{\partial \{(1-\phi)\overline{\rho}_s \overline{C}\}}{\partial t} = \frac{\partial \left[D_B \frac{\partial \{(1-\phi)\overline{\rho}_s \overline{C}\}}{\partial z} \right]}{\partial z} + \frac{\partial \left\{ D'_B (1-\phi)\overline{\rho}_s \frac{\partial \overline{C}}{\partial z} \right\}}{\partial z}$$

$$-\frac{\partial \{(1-\phi)\overline{\rho}_s w\overline{C}\}}{\partial z} + (1-\phi)\overline{\rho}_s \overline{R}_{ads} + (1-\phi)\overline{\rho}_s \sum \overline{R'} \quad (3.2.4)$$

where, $\overline{C}$ = concentration of a particulate substances in terms of mass per unit mass of total solids (mass/mass-solid), $\overline{\rho}_s$ = average density of total solid (mass-solid/ L³-solid), w = rate of depositional burial of solids (L-sediment/T), $\overline{R}_{ads}$ = reactions of dissolved materials due to equilibrium adsorption or desorption (mass/L²-liquid/T), $\sum \overline{R'}$ = all non-equilibrium slow biochemical reactions (mass/mass-solid/T).

In addition to these diagenetic equations, we have the mass balance expression

$$\overline{R}_{ads} = \frac{-\phi}{(1-\phi)\overline{\rho}_s} R_{ads} \quad (3.2.5)$$

Here, following assumptions are made for these equations. (Berner (1980))

I. Average density of total solid does not change with space or time, in other words, $\overline{\rho}_s$ is constant

II. The equilibrium expression for simple linear adsorption: $\overline{C} = K'C$, K' =adsorbed constant.

III. Adsorptive property does not change with space or time, in other words, K' is constant.

IV. If $\overline{C}$ is adsorbed substances, then there are no slow diagenetic reactions, hence, $\sum \overline{R'} = 0$ in equation 3.2.4.

In addition, if $\overline{C}$ is non-adsorbed substances, then $\overline{R}_{ads} = 0$ in equation 3.2.4.

From these equations and assumptions, we can obtain two equations as follows.

For dissolved material

$$\frac{\partial}{\partial t}[\{K\overline{\rho}_s(1-\phi)+\phi\}C] =$$

$$\frac{\partial}{\partial z}\left[\{K\overline{\rho}_s(D_B+D'_B)\cdot(1-\phi)+\phi(D_S+D_I+D_B+D'_B)\}\frac{\partial C}{\partial z}\right]$$

$$+\frac{\partial}{\partial z}\left[\left\{(1-K\overline{\rho}_s)D_B\frac{\partial \phi}{\partial z}-K\overline{\rho}_s(1-\phi)w-\phi v\right\}C\right]$$

$$+\phi\alpha(C_0-C)+\phi\sum R' \qquad (3.2.6)$$

For particulate material

$$\frac{\partial\{(1-\phi)S\}}{\partial t} = \frac{\partial\left\{(D_B+D'_B)\cdot(1-\phi)\frac{\partial S}{\partial x}\right\}}{\partial z}$$

$$+\frac{\partial\left[\left\{D_B\frac{\partial(1-\phi)}{\partial x}-(1-\phi)w\right\}S\right]}{\partial z}+(1-\phi)\overline{\rho}_s\sum \overline{R'} \qquad (3.2.7)$$

where, $\overline{\rho}_s\overline{C} = S$ = concentration of a particulate or adsorbed substances in terms of mass per unit volume of total solids (mass/ L^3-solid).

Equation 3.2.6 is applied to components, dissolved organic matter (labile DOM, refractory DOM), NH_4-N, $NO_{2,3}$-N, PO_4-P, DO, ODU and equation 7 is applied to components, detritus (fast labile POM, slow labile POM, refractory POM).

w and v are pre-calculated to agree with continuity equations (Volume Conservation Equations) as follow.

$$\frac{\partial \phi}{\partial t} + \frac{\partial (v \cdot \phi)}{\partial z} = \frac{\partial}{\partial z}\left(D_B \frac{\partial \phi}{\partial z}\right) \qquad (3.2.8)$$

$$\frac{\partial (1-\phi)}{\partial t} + \frac{\partial \{w \cdot (1-\phi)\}}{\partial z} = \frac{\partial}{\partial x}\left\{D_B \frac{\partial (1-\phi)}{\partial z}\right\} \qquad (3.2.9)$$

From spatial integration of equation 3.2.6 and equation 3.2.7, the model equations can be obtained to use finite volume method (FVM), whose algorithm conserves mass of components. The details are shown in Table 4 and Table 5.

The equations for suspension feeder, deposit feeder, benthic algae, seagrass, seaweed, epiphytes and epifauna are as follows

$$\frac{\partial B}{\partial t} = \sum R_B \qquad (3.2.10)$$

where, B = biomass, expressed per square of sediment (mass/L^2-sediment)
R_B = biochemical reactions (mass/L^2-sediment)

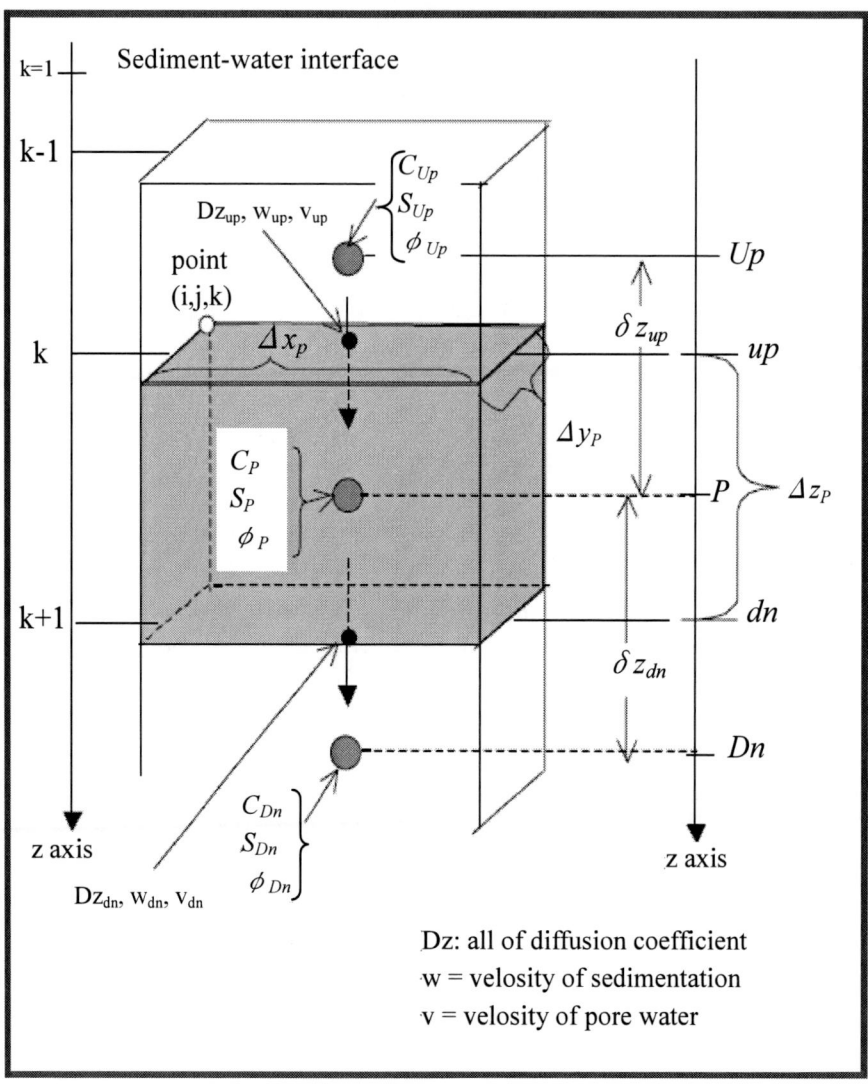

Figure 4. Defined points of variables and coefficients in (i, j, k) mesh/box of benthic system.

Table 4. Model equations for dissolved materials in benthic system (refer to Figure 4)

Dissolved materials: (i,j,k) Mesh/Box
from the spatial integration of (i,j,k) mesh /box,

$$\Delta x \cdot \Delta y \cdot \frac{\partial}{\partial t} \int_{up}^{dn} [\{K'\rho_s(1-\phi) + \phi\}C]dz =$$

$$\Delta x \cdot \Delta y \cdot \left[\{K'\overline{\rho}_s(D_B + D_B')\cdot(1-\phi) + \phi(D_S + D_I + D_B + D_B')\} \frac{\partial C}{\partial z} \right]_{up}^{dn}$$

$$-\Delta x \cdot \Delta y \cdot \left[\left\{ -(1-K'\overline{\rho}_s)D_B \frac{\partial \phi}{\partial z} + K'\rho_s(1-\phi)w + \phi v \right\} C \right]_{up}^{dn}$$

$$+ \Delta x \cdot \Delta y \cdot \int_{up}^{dn} \alpha (C_0 - C)dz + \Delta x \cdot \Delta y \cdot \int_{up}^{dn} (\phi \sum R')dz \quad \text{(T4-1)}$$

$\Leftrightarrow$

$$\frac{\Delta x \cdot \Delta y \cdot \Delta z \cdot (h_P^{t+\Delta t} C_P^{t+\Delta t} - h_P^t C_P^t)}{\Delta t} =$$

$$\Delta x \cdot \Delta y \cdot (F_{up}^{t+\Delta t} - F_{dn}^{t+\Delta t}) + \Delta x \cdot \Delta y \cdot \Delta z \cdot \alpha(C_0 - C_P^t)$$

$$+ \Delta x \cdot \Delta y \cdot \Delta z \cdot \phi_P^t \sum R'_P^t \quad \text{(T4-2)}$$

$\Leftrightarrow$

$$\frac{\Delta x \Delta y \Delta z (h_P^{t+\Delta t} C_P^{t+\Delta t} - h_P^t C_P^t)}{\Delta t} =$$

$$\Delta x \Delta y \cdot (a_{Dn}^{t+\Delta t} C_{Dn}^{t+\Delta t} + a_{Up}^{t+\Delta t} C_{Up}^{t+\Delta t} - a_P^{t+\Delta t} C_P^{t+\Delta t})$$

$$+ \Delta x \Delta y \Delta z \cdot \alpha(C_0 - C_P^t) + \Delta x \Delta y \Delta z \cdot \phi_P^t \sum R'_P^t \quad \text{(T4-3)}$$

$\Leftrightarrow$

$$A_P^{t+\Delta t} C_P^{t+\Delta t} = a_{Dn}^{t+\Delta t} C_{Dn}^{t+\Delta t} + a_{Up}^{t+\Delta t} C_{Up}^{t+\Delta t} + b_P^t \quad \text{(T4-4)}$$

Where, $F_x^t = f_x^t C_x^t - g_x^t \left[\frac{\partial C}{\partial z}\right]_x^t$, $h_z^t = K'\overline{\rho}_s(1-\phi_z^t) + \phi_z^t$,

$$g_z^t = K'\overline{\rho}_s(D_{Bz}^t + D_{Bz}'^t)\cdot(1-\phi_x^t) + \phi_{xz}^t(D_{Sz}^t + D_{Iz}^t + D_{Bz}^t + D_{Bz}'^t),$$

$$f_z^t = -(1-K'\overline{\rho}_s)D_B \left[\frac{\partial \phi}{\partial z}\right]_z^t + K'\rho_s(1-\phi_x^t)w_z^{t-\frac{\Delta t}{2}} + \phi_x^t v_z^{t-\frac{\Delta t}{2}}$$

$$a_{Dn}^t = \max\left[-f_{dn}^t, \frac{g_{dn}^t}{\delta z_{dn}} - \frac{f_{dn}^t}{2}, 0\right], \quad a_{Up}^t = \max\left[f_{up}^t, \frac{g_{up}^t}{\delta z_{up}} + \frac{f_{up}^t}{2}, 0\right],$$

$$a_P^t = a_{Up}^t + a_{Dn}^t + (f_{dn}^t - f_{up}^t)$$

$$A_P^t = a_P^t + \frac{\Delta z}{\Delta t} h_P^t, \quad b_P^t = \frac{\Delta x}{\Delta t} h_P^t C_P^t + \Delta z \cdot \phi_P^t \alpha (C_0 - C_P^t)$$

$$+ \Delta z \cdot \phi_P^t \sum R'_P^t$$

Where the model used the implicit scheme in the vertical direction, and applied the high-lateral flux modification (Spalding 1972, Patankar 1980) both in the vertical and horizontal fluxes. The model solves this equation by Tri-Diagonal Matrix Algorithm (TDMA).

Table 5. Model equations for particulate materials in benthic system (refer to Figure 4)

Particulate materials: (i,j,k) Mesh/Box
From the spatial integration of (i,j,k) mesh /box,

$$\Delta x \Delta y \frac{\partial}{\partial t} \int_{up}^{dn} \{(1-\phi)S\}dz = \Delta x \Delta y \left\{ (D_B + D_B') \cdot (1-\phi) \frac{\partial S}{\partial z} \right\}\Big|_{up}^{dn}$$

$$- \Delta x \Delta y \left[\left\{ -D_B \frac{\partial (1-\phi)}{\partial z} + (1-\phi)w \right\} S \right]\Big|_{up}^{dn} + \Delta x \Delta y \int_{up}^{dn} (1-\phi)\rho_s \sum R' \, dz \quad \text{(T5-1)}$$

$\Leftrightarrow$

$$\frac{\Delta x \Delta y \Delta z \left(hs_P^{t+\Delta t} S_P^{t+\Delta t} - hs_P^{t} S_P^{t} \right)}{\Delta t} = \Delta x \Delta y \left(Fs_{up}^{t+\Delta t} - Fs_{dn}^{t+\Delta t} \right) +$$

$$\Delta x \Delta y \Delta z \cdot (1-\phi_P^t) \overline{\rho}_s \sum \overline{R}'^{t}_P \quad \text{(T5-2)}$$

$\Leftrightarrow$

$$\frac{\Delta x \Delta y \Delta z \left(hs_P^{t+\Delta t} S_P^{t+\Delta t} - hs_P^{t} S_P^{t} \right)}{\Delta t} =$$

$$\Delta x \Delta y \left(as_{Dn}^{t+\Delta t} S_{Dn}^{t+\Delta t} + as_{Up}^{t+\Delta t} S_{Up}^{t+\Delta t} - as_P^{t+\Delta t} S_P^{t+\Delta t} \right)$$

$$+ \Delta x \Delta y \Delta z \cdot (1-\phi_P^t) \overline{\rho}_s \sum \overline{R}'^{t}_P \quad \text{(T5-3)}$$

$\Leftrightarrow$

$$As_P^{t+\Delta t} S_P^{t+\Delta t} = as_{Dn}^{t+\Delta t} S_{Dn}^{t+\Delta t} + as_{Up}^{t+\Delta t} S_{Up}^{t+\Delta t} + bs_P^{t} \quad \text{(T5-4)}$$

Where, $Fs_z^t = fs_z^t S_z^t - gs_z^t \left[\frac{\partial S}{\partial z}\right]_z^t$, $hs_z^t = 1-\phi_z^t$, $gs_z^t = (D_{Bz}^t + D_{Bz}'^t)(1-\phi_z^t)$,

$$fs_z^t = -D_{Bz}^t \left[\frac{\partial (1-\phi)}{\partial z}\right]_z^t + (1-\phi_z^t) w_z^{t-\frac{1}{2}\Delta t,Z}$$

$$as_{Dn}^t = \max\left[-fs_{dn}^t, \frac{gs_{dn}^t}{\delta z_{dn}} - \frac{fs_{dn}^t}{2}, 0\right], \quad as_{Up}^t = \max\left[fs_{up}^t, \frac{gs_{up}^t}{\delta z_{up}} + \frac{fs_{up}^t}{2}, 0\right]$$

$$as_P^t = as_{Up}^t + as_{Dn}^t + (fs_{dn}^t - fs_{up}^t), \quad As_P^t = as_P^t + \frac{\Delta z}{\Delta t} hs_P^t$$

$$bs_P^t = \frac{\Delta z}{\Delta t} hs_P^t S_P^t + \Delta z \cdot (1-\phi_P^t) \overline{\rho}_s \sum \overline{R}'^{t}_P$$

Where the model used the implicit scheme in the vertical direction, and applied the high-lateral flux modification (Spalding 1972, Patankar 1980) both in the vertical and horizontal fluxes. The model solves this equation by Tri-Diagonal Matrix Algorithm (TDMA).

3.2.3. Biochemical Reactions

The major biochemical reactions and rate expressions used in the model are given in Table 6 and Table 7. In the model, most of the biochemical reactions are based on a first kinetics formulation characterized by nonlinear reaction rate terms. Nonlinear reaction rate terms are composed of the functions that represent the limitation of intensity of light in photosynthesis, the limitation or inhibition of model components, and the exponential response to temperature (refer to Tables 6 and 7). Mineralization processes by bacteria are divided into three pathways, though those are indicated as a lump in Figure 2. Those are oxic, suboxic and anoxic mineralization. All these mineralization processes are formulated as first order kinetics of detritus (non-living organic matter) figured with mass of carbon. Oxic mineralization is limited by oxygen (Michaelis-Menten type kinetics). Suboxic mineralization based on nitrate is inhibited by oxygen (one minus Michaelis-Menten type kinetics) and limited by nitrate (Michaelis-Menten type kinetics). The consumption of oxygen and nitrate as terminal electron acceptors is explicitly modeled. Mineralization processes using other oxidants (manganese oxides, iron oxides, sulphate) are lumped into one process, where this mineralization is inhibited by oxygen and nitrate (one minus Michaelis-Menten type kinetics). Anoxic mineralization produces reduced substances called oxygen demand units, ODU. Re-oxidation of one mole of ODU requires one mole of oxygen (Soetaert et al 1996). Part of the ODU is permanently removed as solid substances (e.g. through pyritization or manganese carbonate formation). Idealized stoichiometric relationships of each bacterial mineralization processes that are used in our model are presented in Table 8. We also set the stoichiometric relationships of the consumption/production of organisms by respiration and photosynthesis in the same way as oxic mineralization by bacteria. Detritus (non-living particulate organic matter) are divided into three fractions, fast labile organic matter, slow labile organic matter and refractory organic matter. This means that this model is a Multi-G model. Accordingly, the cells of organisms (phytoplankton, zoo-plankton, sea-grass, seaweed, suspension feeders and deposit feeders, epiphytes, epifauna) are defined as being composed of these three fractions. Hence, fluxes between organisms and detritus (fluxes of uptake, extra-release, feces, mortality, etc) can be modeled by considering these fractions.

Table 6. Formulation of the biochemical processes of a pelagic system

Biochemical Process	Formulation	Unit	Parameters
[Phytoplankton]			
photosynthesis	$Dpp_1 = v_{011} \cdot u_{011} \cdot u_{012} \cdot PP \cdot VolumW$	µgC/hr	
maximum growth rate	$v_{011} = f_{Temp}(\alpha_{11}, \beta_{11})$	1/hr	α_{11}, β_{11}
nutrient limitation	$u_{011} = \min(g(WNX + WNY, Hf_{n01}), g(WDP, Hf_{p01}))$	-	Hf_{n01}, Hf_{p01}
light availability	$u_{012} = \dfrac{1}{\Delta z} \int_{z}^{z+\Delta z} \dfrac{I_0}{I_{opt}} e^{-kz} \exp\left\{1 - \dfrac{I_0}{I_{opt}} e^{-kz}\right\} dz$		I_{opt}
	$z, \Delta z$: depth and thickness of the layer	cm	
light attenuation	$k = k_{01} + \gamma_{PP1} \cdot PP_{av} + \gamma_{ZP1} \cdot ZP_{av} + \gamma_{POC1} \cdot (WFP_{av} + WSP_{av} + WGP_{av})$ (in high tide)	1/cm	$k_{01}, \gamma_{PP1}, \gamma_{ZP1}, \gamma_{POC1}$
	$k_{02} + \gamma_{PP2} \cdot PP_{av} + \gamma_{ZP2} \cdot ZP_{av} + \gamma_{POC2} \cdot (WFP_{av} + WSP_{av} + WGP_{av})$ (in low tide)	1/cm	$k_{02}, \gamma_{PP2}, \gamma_{ZP2}, \gamma_{POC2}$
	$PP_{av}, ZP_{av}, WFP_{av}, WSP_{av}, WGP_{av}$: average of PP, ZP, WFP, WSP, WGP in water column.	mgC/l	
surface light intensity	$I_0(t)$: prescribed function	µE/m²/sec	
extra-release	$Dpp_2 = Dpp_1 \cdot 0.135 \cdot \exp(-0.00201 \cdot Rgrm01 \cdot PP \cdot 10^3)$	µgC/hr	Rgrm01
respiration	$Dpp_3 = f_{Temp}(\alpha_{12}, \beta_{12}) \cdot PP \cdot VolumW$	µgC/hr	α_{12}, β_{12}
feeding by zooplankton	$Dpp_4 = Dzp_1$	µgC/hr	
natural mortality	$Dpp_5 = f_{Temp}(\alpha_{13}, \beta_{13}) \cdot PP \cdot VolumW$	µgC/hr	α_{13}, β_{13}
[Zooplankton]			
feeding	$Dzp_1 = f_{Temp}(\alpha_{21}, \beta_{21}) \cdot (1 - \exp(\lambda \cdot (\Pi - PP))) \cdot ZP \cdot VolumW$	µgC/hr	$\alpha_{21}, \beta_{21}, \lambda, \Pi$
feces	$Dzp_2 = (1-e) \cdot Dzp_1$	µgC/hr	e
excretion	$Dzp_3 = (e-g) \cdot Dzp_1$	µgC/hr	e, g
mortality	$Dzp_4 = f_{Temp}(\alpha_{22}, \beta_{22}) \cdot ZP \cdot VolumW$	µgC/hr	α_{22}, β_{22}
[Detritus, Dissolved organic matter]			
mineralization	$Min_i = OxicMin_i + SuboxicMin_i + AnoxicMin_i$	µgC/hr	
oxic mineralization	$OxicMin_i = R_{wi} \cdot g(WDO, Kw_{sO2}) \cdot TOC_i \cdot VolumW / \Sigma$	µgC/hr	Kw_{sO2}

Table 6. (Continued).

Biochemical Process	Formulation	Unit	Parameters
suboxic mineralization	$SuboxicMin_i = R_{wi} \cdot g(WHY, Kw^{Sub}_{s\,NO3})$ $\cdot h(WDO, Kw^{Sub}_{in\,O2}) \cdot TOC_i \cdot VolumW / \Sigma$	µgC/hr	$Kw^{Sub}_{in\,O2}$, $Kw^{Sub}_{s\,NO3}$
anoxic mineralization	$AnoxicMin_i = R_{wi} \cdot h(WHY, Kw^{An}_{in\,NO3})$ $\cdot h(WDO, Kw^{An}_{in\,O2}) \cdot TOC_i \cdot VolumW / \Sigma$	µgC/hr	$Kw^{An}_{in\,O2}$, $Kw^{An}_{in\,NO3}$
	$\Sigma = g(WDO, Kw_{sO2})$ $+ g(WHY, Kw^{Sub}_{s\,NO3}) \cdot h(WDO, Kw^{Sub}_{in\,O2})$ $+ h(WHY, Kw^{An}_{in\,NO3}) \cdot h(WDO, Kw^{An}_{in\,O2})$	-	Kw_{sO2}, $Kw^{Sub}_{in\,O2}$, $Kw^{Sub}_{s\,NO3}$, $Kw^{An}_{in\,O2}$, $Kw^{An}_{in\,NO3}$
	TOC_i, $i=3,4,5,6,7$ for WFP, WSP, WGP, WBM, WGM.	mgC/l	
Maximum mineralization rate	$R_{wi} = f_{Temp}(\alpha_{i1}, \beta_{i1})$ $i=3,4,5,6,7$ for WFP, WSP, WGP, WBM, WGM	1/hr	$\alpha_{31}, \beta_{31}, \alpha_{41}, \beta_{41},$ α_{51}, β_{51} $\alpha_{61}, \beta_{61}, \alpha_{71}, \beta_{71}$
bio-degradation of WFP	$Dwfp_4 = R_{min03} \cdot Min_3$	µgC/hr	R_{min03}
bio-degradation of WSP	$Dwsp_4 = R_{min04} \cdot Min_4$	µgC/hr	R_{min04}
bio-degradation of WGP	$Dwgp_4 = R_{min05} \cdot Min_5$	µgC/hr	R_{min05}
[NH$_3$-N, NO$_{2,3}$-N]			
nitrification	$Dwnx_1 = f_{Temp}(\alpha_{81}, \beta_{81}) \cdot g(WDO, Hf_{wno1}) \cdot WNX \cdot VolumW$	µgN/hr	$\alpha_{81}, \beta_{81}, Hf_{wno1}$
nitrate reduction	$Dwnx_{10} = (14/12) \cdot (4/(8-3 \cdot A_{wyy})) \cdot (1 - A_{wyy}) \cdot \sum_{i=3}^{7} SuboxicMin_i$	µgN/hr	A_{wyy}
denitrification	$Dwny_2 = (14/12) \cdot (4/(8-3 \cdot A_{wyy})) \cdot A_{wyy} \cdot \sum_{i=3}^{7} SuboxicMin_i$	µgN/hr	A_{wyy}
[ODU]			
production	$Dwou_1 = (32/12) \cdot \sum_{i=3}^{7} AnoxicMin_i$	µgODU/hr	
oxidation	$Dwou_2 = f_{Temp}(\alpha_{111}, \beta_{111}) \cdot g(WDO, Hf_{wouo}) \cdot WOU \cdot VolumW$ $+ A_{sanka} \cdot Dwou_1$	µgODU/hr	$\alpha_{111}, \beta_{111}$, Hf_{wouo}, A_{sanka}
Biochemical Process	Formulation	Unit	Parameters

Table 6. (Continued).

			µgODU/hr	A_{wauth}, A_{waut2}
authigenic mineralization	$Dwou_3 = A_{wauth} \cdot Dwou_2 + A_{waut2} \cdot Dwou_1 + f_{Temp}(\alpha_{112}, \beta_{112}) \cdot WOU \cdot VolumW$			
[Eelgrass]				
photosynthesis	$Delg_1 = v_{131} \cdot u_{131} \cdot u_{132} \cdot R_{gw13} \cdot ELG \cdot Squre$		µgC/hr	R_{gw13}
maximum growth rate	$v_{131} = \alpha_{131} \cdot (1 - A_{sl131} \cdot (TpBt - T_{opt13})^2)$ (tpBt $\leq T_{opt13}$)		1/hr	α_{131}, T_{opt13}, A_{sl131}, A_{sl132}
	$\alpha_{131} \cdot (1 - A_{sl132} \cdot (TpBt - T_{opt13})^2)$ (tpBt $> T_{opt13}$)			
	tpBt: temperature of sediment		degree	
nutrient limitation	$u_{131} = \min(g(\max(HNX_{av} + HNY_{av}, WNX_{av} + WNY_{av}), Hf_{n13}),$		-	Hf_{n13}, Hf_{p13}
	$g(\max(DIP_{av}, WDP_{av}), Hf_{p13}))$			
	HNX_{av}, HNY_{av}, WNX_{av}, WNY_{av}, DIP_{av}, WDP_{av} :		mg/l	
	average of HNX,HNY,WNX,WNY,DIP,WDP in column.			
light availability	$u_{132} = I_0 e^{-k \cdot (depth-50.0)} / (I_{opt13} + I_0 e^{-k \cdot (depth-50.0)})$		-	I_{opt13}
	depth : water depth		cm	
extra-release	$Delg_2 = A_{1306} \cdot Delg_1$		µgC/hr	A_{1306}
respiration	$Delg_3 = A_{13081} \cdot v_{131} \cdot ELG \cdot Squre + A_{13082} \cdot Delg_1$		µgC/hr	A_{13081}, A_{13082}
natural mortality	$Delg_4 = f_{TpBt}(\alpha_{132}, \beta_{132}) \cdot ELG \cdot Squre$		µgC/hr	α_{132}, β_{132}
[Ulva]				
photosynthesis	$Dulv_1 = v_{141} \cdot u_{141} \cdot u_{142} \cdot ULV \cdot Squre$		µgC/hr	
maximum growth rate	$v_{141} = \alpha_{141} \cdot (1 - A_{sl141} \cdot (TpBt - T_{opt14})^2)$ (tpBt $\leq T_{opt14}$)		1/hr	α_{141}, T_{opt14}, A_{sl141}, A_{sl142}
	$\alpha_{141} \cdot (1 - A_{sl142} \cdot (TpBt - T_{opt14})^2)$ (tpBt $> T_{opt14}$)			
nutrient limitation	$u_{141} = \min(g(WNX_{av} + WNY_{av}, Hf_{n14}), g(WDP_{av}, Hf_{p14}))$		-	Hf_{n14}, Hf_{p14}
light availability	$u_{142} = \frac{1}{depth} \int_0^{depth} \frac{I_0}{I_{opt14}} e^{-kz} \exp\left[1 - \frac{I_0}{I_{opt14}} e^{-kz}\right] dz$		-	I_{opt14}
extra-release	$Dulv_2 = A_{1406} \cdot Dulv_1$		µgC/hr	A_{1406}
respiration	$Dulv_3 = A_{14081} \cdot v_{141} \cdot ULV \cdot Squre + A_{14082} \cdot Dulv_1$		µgC/hr	A_{14081}, A_{14082}
natural mortality	$Dulv_4 = f_{TpBt}(\alpha_{142}, \beta_{142}) \cdot ULV \cdot Squre$		µgC/hr	α_{142}, β_{142}
[Epiphyte]				
photosynthesis	$Dzpp_1 = v_{161} \cdot u_{161} \cdot u_{162} \cdot Rz_{16} \cdot ZPP \cdot Squlef$		µgC/hr	

Table 6. (Continued).

Biochemical Process	Formulation	Unit	Parameters
maximum growth rate	$v_{161} = f_{Temp}(\alpha_{161}, \beta_{161})$	1/hr	$\alpha_{161}, \beta_{161}$
nutrient limitation	$u_{161} = \min(g(WNX+WNY, Hf_{n161}), g(WDP, Hf_{p161}))$	-	Hf_{n161}, Hf_{p161}
light availability	$u_{162} = \dfrac{1}{\Delta z}\int_{z}^{z+\Delta z}\dfrac{I_0}{I_{opt16}}e^{-kz}\exp\left\{1-\dfrac{I_0}{I_{opt16}}e^{-kz}\right\}dz$	-	I_{opt16}
vertical distribution ratio	Rz_{16} = (thickness of layer)/(thickness of water column)	-	
extra-release	$Dzpp_2 = Dzpp_1 \cdot 0.135 \cdot \exp(-0.00201 \cdot R_{grm01} \cdot Rz_{16} \cdot ZPP \cdot 10^3)$	µgC/hr	R_{grm01}
respiration	$Dzpp_3 = f_{Temp}(\alpha_{162}, \beta_{162}) \cdot Rz_{16} \cdot ZPP \cdot Squlef$	µgC/hr	$\alpha_{162}, \beta_{162}$
feeding by epifauna	$Dzpp_4 = f_{Temp}(\alpha_{171}, \beta_{171}) \cdot (1-\exp(\lambda_{17} \cdot (\Pi_{17} - Rz_{16} \cdot ZPP))) \cdot Rz_{17} \cdot ZZP \cdot Squlef$	µgC/hr	$\alpha_{171}, \beta_{171}, \lambda_{17}, \Pi_{17}$
natural mortality	$Dzpp_5 = f_{Temp}(\alpha_{163}, \beta_{163}) \cdot Rz_{16} \cdot ZPP \cdot Squlef$	µgC/hr	$\alpha_{163}, \beta_{163}$
[Epifauna]			
feeding	$Dzzp_1 = Dzpp_4$	µgC/hr	
feces	$Dzzp_2 = (1-e_{17}) \cdot Dzzp_1$	µgC/hr	e_{17}
excretion	$Dzzp_3 = (e_{17}-g_{17}) \cdot Dzzp_1$	µgC/hr	e_{17}, g_{17}
mortality	$Dzzp_4 = f_{Temp}(\alpha_{172}, \beta_{172}) \cdot Rz_{17} \cdot ZZP \cdot Squlef$	µgC/hr	$\alpha_{172}, \beta_{172}$
vertical distribution ratio	Rz_{17} = (thickness of layer) / (thickness of water column)	-	

functions : $f_{Temp}(\alpha, \beta) = \alpha \exp(\beta \cdot Temp)$ $\min(a, b) = \begin{cases} a & (a \leq b) \\ b & (a > b) \end{cases}$ $g(X, a_{half}) = \dfrac{X}{X+a_{half}}$ $h(X, a_{half}) = (1 - \dfrac{X}{X+a_{half}})$.

Mesh informations : VolumW = volume of mesh (cm^3) Squre = area of mesh (cm^2) Squlef = area of leaf (prescribed function) (cm^2).

Model component	Units	Description
PP	mgC/l	Phytoplankton
ZP	mgC/l	Zooplankton
WFP, WSP, WGP	mgC/l	Detritus (fast-labile POM, slow-labile POM, refractory POM)
WBM, WGM	mgC/l	Dissolved organic carbon (labile DOM, refractory DOM)
WNX	mgN/l	NH_4-N
WNY	mgN/l	$NO_{2,3}$-N
WDP	mgP/l	PO_4-P
WOU	mg/l	ODU
WDO	mg/l	Dissolved oxygen
ELG	µgC/cm^2 sed.	Eelgrass
ULV	µgC/cm^2 sed.	Ulva
ZPP	µgC/cm^2 leaf	Epiphyte
ZZP	µgC/cm^2 leaf	Epifauna

Table 7. Formulation of the biochemical processes of a benthic system

Biochemical Process	Formulation	Unit	Parameters
[Suspension feeder]			
Feeding	$Dsfb_1 = \min(Dsfb_{1b}, eatmax)$	μgC/hr	
filter rate limitation	$Dsfb_{1b} = v_{511} \cdot u_{511} \cdot (PP + ZP + WFP + WSP + WGP + WBM + WGM) \cdot R_{cor0} \cdot SFB \cdot Squre$	μgC/hr	
filter rate	$v_{511} = 1.2 \cdot 10^{-5} \cdot tpBt^{1.25} \cdot Wwet_{51}^{-0.75} / Rwd_{51} / Rcd_{51}$ (tpBt≥10) $1.2 \cdot 10^{-5} \cdot 10^{1.25} \cdot Wwet_{51}^{-0.75} / Rwd_{51} / Rcd_{51}$ (tpBt<10)	ml/hr/μgC	$Wwet_{51}$, Rwd_{51}, Rcd_{51}
oxygen saturation limitation	$u_{511} = \min(1, dorat/deDO_{51})$ dorat : oxygen saturation of bottom layer of pelagic system (calculated)	-	$deDO_{51}$
decreasing ratio by double filtering	$R_{cor0} = (1 - \exp(-COR0))/COR0$	-	
double filtering ratio	$COR0 = v_{511} \cdot u_{511} \cdot SFB \cdot dt/dzzSF$	-	dt, $dzzSF$
growth rate limitation	$eatmax = (AgrtSF + AbasSF)/(Aege_{51} \cdot (1 - Aexc_{51})) \cdot SFB \cdot F_{temp} \cdot Squre$ F_{temp}: function of temperature	μgC/hr	$AgrtSF$, $AbasSF$, $Aege_{51}$, $Aexc_{51}$
Feces	$Dsfb_2 = (1 - Aege_{51}) \cdot Dsfb_1$	μgC/hr	
Excretion	$Dsfb_3 = Aexc_{51} \cdot Aege_{51} \cdot Dsfb_1 + AbasSF \cdot SFB \cdot Squre$	μgC/hr	
Mortality	$Dsfb_4 = u_{512} \cdot v_{512} \cdot SFB \cdot Squre$	μgC/hr	
temperature dependency	$v_{512} = f_{tpBt}(1, \beta_{512})$	-	β_{512}
rate of mortality	$u_{512} = Ade_{511} + Ade_{512} \cdot (1 - u_{511})$	1/hr	Ade_{511}, Ade_{512}
[Deposit feeder]			
Feeding	$Ddfb_1 = v_{521} \cdot u_{521} \cdot u_{522} \cdot u_{523} \cdot DFB \cdot Squre$	μgC/hr	

Table 7. (Continued).

Biochemical Process	Formulation	Unit	Parameters
maximum ration	$v_{521}=f_{tpBt}(\alpha_{521}, \beta_{521})$	1/hr	$\alpha_{521}, \beta_{521},$ $Teat_{52}$
food limitation	$u_{521}=1-\exp(Aivl_{52} \cdot \min(0, Akai_{52}-Food_{52}))$ $Food_{52}$: average concentration of food in mud. (calculated)	- $\mu gC/cm^3$	$Aivl_{52}, Akai_{52}$
cannibalism efficiency	$u_{522}=g(DFB, Hf_{ea52})$	$\mu gC/cm^2$	Hf_{ea52}
oxygen saturation limitation	$u_{523}=\min(1, dorat/deDO_{52})$	-	$deDO_{52}$
Feces	$Ddfb_2=(1-u_{524}) \cdot Ddfb_1$	$\mu gC/hr$	
assimilation efficiency	$u_{524}=1-Andg_{52} \cdot (1+g(Food_{52}, Hf_{du52}))$	-	$Andg_{52}, Hf_{du52}$
Excretion	$Ddfb_3=Aexc_{52} \cdot u_{524} \cdot Ddfb_1$	$\mu gC/hr$	$Aexc_{52}$
Mortality	$Ddfb_4=v_{522} \cdot u_{525} \cdot DFB \cdot Squre$	$\mu gC/hr$	
temperature dependency	$v_{522}=\min(f_{tpBt}(1, \beta_{521}), f_{Teat52}(1, \beta_{521}))$	-	β_{522}
rate of mortality	$u_{525}=Ade_{521}+Ade_{522} \cdot (1-u_{523})$	1/hr	Ade_{521}, Ade_{522}
[Detritus, Dissolved organic carbon]			
mineralization of Detritus	$Min_{POM}=Min_3+Min_4+Min_5+Min_8+Min_9$	$\mu gC/hr$	
mineralization of DOM	$Min_{DOM}=Min_6+Min_7$ $i=3\sim 9$ for FBP,SBP,GPM,BMM,GMM,adhered BMM, adhered GMM	$\mu gC/hr$	
mineralization of component	$Min_i=OxicMin_i + SuboxicMin_i + AnoxicMin_i$ 1st layer $SuboxicMin_i + AnoxicMin_i$ other layers	$\mu gC/hr$	
oxic mineralization	$OxicMin_i=R_{bi} \cdot g(oxicB, Ko_{sO2}) \cdot TOCb_i \cdot (Io/I) / \Sigma o$	$\mu gC/hr$	l, Ko_{sO2}
suboxic mineralization	$SuboxicMin_i=(Io/I) \cdot SuboxicMinO_i + ((l-Io)/I) \cdot SuboxicMinB_i$ 1st layer $SuboxicMinB_i$ other layers	$\mu gC/hr$	1

Table 7. (Continued).

Biochemical Process	Formulation	Unit	Parameters
oxic layer	SuboxicMinO$_l$=R$_{bl}$·g(HNY, Ko$^{Sub}_{s\ NO3}$)·h(oxicB, Ko$^{Sub}_{in\ O2}$)·TOCb$_l$/Σo	μgC/hr	Ko$^{Sub}_{in\ O2}$, Ko$^{Sub}_{s\ NO3}$

Biochemical Process	Formulation	Unit	Parameters
anoxic layer	SuboxicMinB$_l$=R$_{bl}$·g(HNY, Kb$^{Sub}_{s\ NO3}$)·TOCb$_l$/Σb	μgC/hr	Kb$^{Sub}_{s\ NO3}$
anoxic mineralization	AnoxicMin$_l$=(lo/l)·AnoxicMinO$_l$ + ((l-lo)/l)·AnoxicMinB$_l$, 1st layer AnoxicMinB$_l$, other layers	μgC/hr	1
oxic layer	AnoxicMinO$_l$=R$_{bl}$·h(HNY, Ko$^{An}_{in\ NO3}$)·h(oxicB, Ko$^{An}_{in\ O2}$)·TOCb$_l$/Σo	μgC/hr	Ko$^{An}_{in\ O2}$, Ko$^{An}_{in\ NO3}$
anoxic layer	AnoxicMinB$_l$=R$_{bl}$·h(HNY, Kb$^{An}_{in\ NO3}$)·TOCb$_l$/Σb	μgC/hr	Kb$^{An}_{in\ NO3}$
	TOCb$_3$=FBP·VolumB·(1-φ)	μgC	
	TOCb$_4$=SBP·VolumB·(1-φ)	μgC	
	TOCb$_5$=GPM·VolumB·(1-φ)	μgC	
	TOCb$_6$=BMM·VolumB·φ	μgC	
	TOCb$_7$=GMM·VolumB·φ	μgC	
	TOCb$_8$=BMM·VolumB·(1-φ)·ρ·K$_{BMM}$	μgC	ρ, K$_{BMM}$, K$_{GMM}$
	TOCb$_9$=GMM·VolumB·(1-φ)·ρ·K$_{GMM}$	μgC	ρ, K$_{BMM}$, K$_{GMM}$
	Σo= g(oxicB, Ko$_{sO2}$)+g(HNY, Ko$^{Sub}_{s\ NO3}$)·h(oxicB, Ko$^{Sub}_{in\ O2}$) +h(HNY, Ko$^{An}_{in\ NO3}$)·h(oxicB, Ko$^{An}_{in\ O2}$)	-	Ko$_{sO2}$, Ko$^{Sub}_{inO2}$, Ko$^{An}_{inO2}$, Ko$^{Sub}_{s\ NO3}$, Ko$^{An}_{s\ NO3}$, Ko$^{An}_{in\ NO3}$
	Σb= g(HNY, Kb$^{Sub}_{s\ NO3}$) + h(HNY, Kb$^{An}_{in\ NO3}$)	-	Kb$^{Sub}_{s\ NO3}$, Kb$^{An}_{in\ NO3}$

Table 7. (Continued).

Biochemical Process	Formulation	Unit	Parameters
	OxicB : concentration of O_2 in oxic layer. (calculated)	mgO_2/l	
	lo : thickness of oxic layer. (calculated)	cm	
Maximum mineralization rate	$R_{bi} = f_{tpBt}(\alpha_{5i1}, \beta_{5i1})$	1/hr	$\alpha_{531}, \beta_{531}, \alpha_{541}, \beta_{541}, \alpha_{551}, \beta_{551}, \beta_{561}, \alpha_{571}, \beta_{571}$
	$i=3,4,5,6,7$ for FBP, SBP, GPM, BMM, GMM. $R_{b8} = R_{b6}, R_{b9} = R_{b7}$		$\alpha_{561}, \beta_{561}, \alpha_{571}, \beta_{571}$
bio-degradation of FBP	$Dfbp_4 = R_{min53} \cdot Min_3$	$\mu gC/hr$	R_{min53}
bio-degradation of SBP	$Dsbp_4 = R_{min54} \cdot Min_4$	$\mu gC/hr$	R_{min54}
bio-degradation of GPM	$Dgpm_4 = R_{min55} \cdot Min_5$	$\mu gC/hr$	R_{min55}
[NH$_4$-N, NO$_{2,3}$-N]			
Nitrification	$Dhnx_1 = f_{tpBt}(\alpha_{581}, \beta_{581}) \cdot g(oxicB, Hf_{mio1}) \cdot HNX \cdot (lo/l) \cdot VolumB \cdot \phi$	$\mu gN/hr$	$\alpha_{581}, \beta_{581}, Hf_{mio1}$
nitrate reduction	$Dhnx_9 = (14/12) \cdot (4/(8 \cdot 3 \cdot A_{yyy})) \cdot (1-A_{yyy}) \cdot \sum_{i=3}^{9} SuboxicMin_i$	$\mu gN/hr$	A_{yyy}
Denitrification	$Dhny_2 = (14/12) \cdot (4/(8 \cdot 3 \cdot A_{yyy})) \cdot A_{yyy} \cdot \sum_{i=3}^{9} SuboxicMin_i$	$\mu gN/hr$	A_{yyy}
[ODU]			
Production	$Dodu_1 = (32/12) \cdot \sum_{i=3}^{9} AnoxicMin_i$	$\mu gODU/hr$	
Oxidation	$Dodu_2 = f_{tpBt}(\alpha_{611}, \beta_{611}) \cdot g(oxicB, Hf_{oduo}) \cdot ODU \cdot (lo/l) \cdot VolumB \cdot \phi + A_{sanka} \cdot Dodu_1$ 1st layer (surface) $A_{sanka} \cdot Dodu_1$ other layers	$\mu gODU/hr$	$\alpha_{611}, \beta_{611}, Hf_{oduo}, A_{sanka}$
authigenic mineralization	$Dodu_3 = A_{authg} \cdot Dodu_2 + A_{auth2} \cdot Dodu_1$ $+ f_{tpBt}(\alpha_{612}, \beta_{612}) \cdot ODU \cdot VolumB \cdot \phi$	$\mu gODU/hr$	$A_{authg}, \alpha_{612}, \beta_{612}$

Table 7. (Continued).

Biochemical Process	Formulation	Unit	Parameters
[Benthic algae]			
Photosynthesis	$Ddia_1 = v_{631} \cdot u_{631} \cdot u_{632} \cdot fbdia \cdot Squre$	μgC/hr	
maximum growth rate	$v_{631} = f_{tpBt}(\alpha_{631}, \beta_{631})$	1/hr	$\alpha_{631}, \beta_{631}$
nutrient limitation	$u_{631} = \min(g(HNX+HNY, Hf_{n63}), g(DIP, Hf_{p63}))$	-	Hf_{n63}, Hf_{p63}
light limitation	$u_{632} = I_0 e^{-Pref63 \cdot depth}/(I_{opt63} + I_0 e^{-Pref63 \cdot depth})$	-	I_{opt63}
light attenuation	$Pref_{63}$: calculated	1/cm	$Prf_{011}, Prf_{012}, Prf_{631}$
light available biomass	$Fbdia = (2/3.14) \cdot DIA \cdot \arctan(Zph_{63}/CBO)$	μgC/cm²	CBO
light available thickness	Zph_{63}: depend on space	cm	
extra-release	$Ddia_2 = A_{6356} \cdot Ddia_1$	μgC/hr	A_{6356}
Excretion	$Ddia_3 = A_{63581} \cdot v_{631} \cdot fbdia \cdot Squre + A_{63582} \cdot Ddia_1$	μgC/hr	A_{63581}, A_{63582}
Mortality	$Ddia_4 = f_{tpBt}(\alpha_{632}, \beta_{632}) \cdot DIA \cdot Squre$	μgC/hr	$\alpha_{632}, \beta_{632}$

functions : $f_{tpBt}(\alpha,\beta) = \alpha \exp(\beta \cdot tpBt)$ $\min(a, b) = \begin{cases} a & (a \leq b) \\ b & (a > b) \end{cases}$ $g(X, a_{half}) = \dfrac{X}{X+a_{half}}$ $h(X, a_{half}) = (1 - \dfrac{X}{X+a_{half}})$

Mesh informations : VolumB = volume of mesh (cm³) Squre = area of mesh (cm²) φ = porosity (-)

Model component	Units	Description
SFB	μgC/cm²(sed.)	Suspension feeder
DFB	μgC/cm²(sed.)	Deposit feeder
FBP, SBP, GPM	μgC/cm³(solid)	Detritus (fast-labile POM, slow-labile POM, refractory POM)
BMM, GMM	MgC/l	Dissolved organic carbon (labile DOM, refractory DOM)
HNX	MgN/l	NH_4-N
HNY	MgN/l	$NO_{2,3}$-N
DIP	mgP/l	PO_4-P
ODU	mg/l	ODU
DIA	μgC/cm²(sed.)	Benthic algae

Table 8. Stoichiometric relationships associated with biochemical processes treated in the ecological model

Photosynthesis using NH_4-N	
$m(CO_2) + n(NH_3) + (H_3PO_4) + m(H_2O) \rightarrow (CH_2O)_m(NH_3)_n(H_3PO_4) + m(O_2)$	(T8.1)
Photosynthesis using NO_3-N	
$m(CO_2) + n(NO_3^-) + (H_3PO_4) + (m+n)(H_2O) + nH^+ \rightarrow (CH_2O)_m(NH_3)_n(H_3PO_4) + (2n+m)(O_2)$	(T8.2)
Oxic mineralization, Excretion, Respiration	
$(CH_2O)_m(NH_3)_n(H_3PO_4) + m(O_2) \rightarrow m(CO_2) + n(NH_3) + (H_3PO_4) + m(H_2O)$	(T8.3)
Suboxic mineralization	
$(CH_2O)_m(NH_3)_n(H_3PO_4) + a(HNO_3) \rightarrow m(CO_2) + a(x/2)(N_2) + n(NH_3) + a(1-x)(NH_3) + (H_3PO_4) + b(H_2O)$	(T8.4)
where, $a = -4m/(3x-8)$, $b = m(3x-4)/(3x-8)$, $0 \leq x \leq 1$, These condition satisfy $a \geq 0$ and $b \geq 0$ at anytime	
Anoxic mineralization	
$(CH_2O)_m(NH_3)_n(H_3PO_4) + m(TEA) \rightarrow m(CO_2) + n(NH_3) + (H_3PO_4) + m(ODU) + Q(H_2O)$	(T8.5)
where, $ODU = 2Mn^{2+}$, $4Fe^{2+}$ and $1/2 S^{2-}$, $TEA = 2 MnO_2$, $2 Fe_2O_3$ and $1/2 SO_4^{2-}$	
Nitrification	
$NH_3 + H_2O + 2 O_2 \rightarrow NO_3^- + 2 H_2O + H^+$	(T8.6)
ODU (Oxygen Demand Unit) oxidization	
$ODU + O_2 \rightarrow TEA$	
where, $ODU = 2 Mn^{2+}$, $4 Fe^{2+}$ and $1/2 S^{2-}$, $TEA = 2 MnO_2$, $2 Fe_2O_3$ and $1/2 SO_4^{2-}$	(T8.7)

(1) m, n denote C, N, P ratio of created or mineralized organic matter, i.e., C:N:P=m:n:1. (2) x denotes ratio of nitrogen reducing to nitrogen gas (N_2) and reducing to ammonium from nitrate by suboxic mineralization. i.e., N_2:NH_3= x:(1-x). (3) a, b are coefficients determined from stoichiometric relation. (4) Nitrogen of N_2 and NH_3 shading and written by italic in the right-hand side are derived from HNO_3 in the left-hand side in the equation "T8.4".

3.2.4. Dissolved Oxygen in Benthic System

Treatment of dissolved oxygen in the benthic system is different from other model components, though dissolved oxygen in the pelagic system is treated with the same equation and schemes of other pelagic components. In the benthic system, the oxygen layer and the anoxic layer are defined. Vertical construction of meshes/boxes is described in Figure. 5. The 1st layer (surface layer) of benthic system consists of an oxic layer and an anoxic layer. The coordinates of the boundary point between oxic and anoxic layer are decided from the total oxygen consumption and production in benthic system, which are calculated from biochemical fluxes at every time step. The way to solve the thickness of the oxic layer in benthic system is given in Table 9.

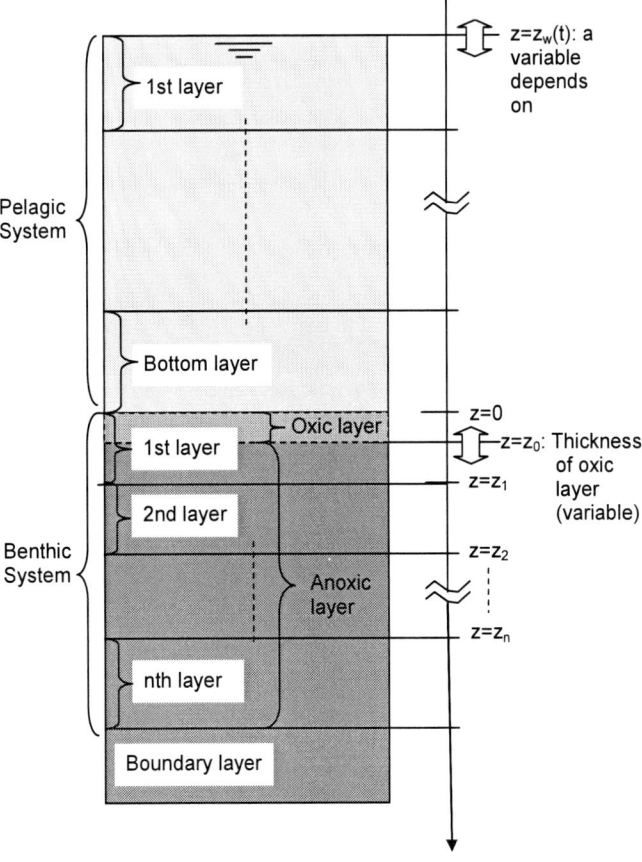

Figure 5. Vertical construction of mesh.

Table 9. Treatment of dissolved oxygen in the benthic system

Temporal and spatial distribution of dissolved oxygen is as follows

$$\frac{\partial DO}{\partial t} = \frac{\partial}{\partial z}\left(D_s \frac{\partial DO}{\partial z}\right) + \sum P_{benthic} \quad \text{(T9-1)}$$

Assumption 1 : $\frac{\partial DO}{\partial t} = 0$, Assumption 2 : $D_s \frac{\partial DO}{\partial z}\bigg|_{z=z_0} = F_{ODU}$,

Assumption 3 : $DO(0) = DO_0$

From these assumptions, equation in benthic system can be solved as given below

$$DO(Z) = \frac{-\sum P_{ebnthic}}{2D_s}Z^2 + \frac{\sum P_{benthic}Z_0 + F_{ODU}}{D_s}Z + DO_0 \quad \text{(T9-2)}$$

$$Z_0 = \frac{-\frac{F_{ODU}}{D_s} - \sqrt{\left(\frac{F_{ODU}}{D_s}\right)^2 - \frac{2\sum P_{benthic}}{D_s}DO_0}}{\left(-\frac{\sum P}{D_s}\right)} \quad \text{(T9-3)}$$

In the oxic layer

$\sum P_{benthic}$ = (production by benthic algae photosynthesis)
- (consumption by oxic mineralization of detritus)
- (consumption by nitrification in benthic system)
- (consumption by oxidation of ODU (oxygen demand materials))

Oxygen production is calculated from biochemical fluxes to use the stoichiometric relations of model components which are given in Table 8. Oxygen production from macro benthos respiration are considered at the bottom layer of pelagic system.

$DO(mgO_2/l)$: Concentration of dissolved oxygen,

$D_z(cm^2/day)$: Diffusion coefficient of dissolved oxygen in the pore water

$\sum P_{benthic}$ (mgO_2/l): Oxygen production in the benthic system

F_{ODU} (mgO_2/cm^2): Oxygen consumption by ODU existed in the anoxic layer (negative value)

Z_0(cm): Thickness of the oxic layer

Chapter 4

IMPLEMENTATION

In this section, how to apply the ecosystem model, TRÄUMEREI to the Jinno area (Figure. 1) is presented.

4.1. EXPRESSION OF THE GEOGRAPHICAL FEATURE

When the ecosystem model, TRÄUMEREI is applied to the Jinno area of Atsumi Bay, the hydrodynamic model is also calculated (refer to *"3.1. Whole Construction of Simulation System"*).

On the hydrodynamics model simulation, first, we started to calculate Ise Bay including Atsumi Bay (Figure 1), because the current field of Atsumi Bay is much affected by the whole Ise Bay. Though fine resolution was desirable to account for the complicated estuarine topography, this simulation was calculated by 2km*2km mesh in horizontal directions and divided into 6 different-sized layers in vertical directions. However, this spatial horizontal resolution is insufficient to estimate the differences of characteristic parts of the Jinno area in Atsumi Bay. Therefore, we took the time series results of Ise Bay simulation as open boundary conditions of Atsumi Bay, and recalculated Atsumi Bay with high resolution around the Jinno area.

The ecosystem model (the TRÄUMEREI model) simulation was applied to meshes around the Jinno area (Figure 1). The depth, the number of pelagic layers in the vertical direction and location of field surveys, which were carried out from April 2000 to February 2001, are also shown in Figure 1. In the vertical direction, the water column was divided into 1 m slices without surface mesh = 2 m. The sediment column was divided into 3 different-sized slices. We set the surface layer = 1cm, middle layer = 1cm, bottom layer = 8 cm. Sea-grass beds (eelgrass beds) are located at the coordinates of the mesh points, (i, j) = (4, 4), (4, 5). Submerged/emerged area is located in (4, 3) and fresh water flow from Toyokawa River comes from (2, 5) in Figure 1.

4.2. BOUNDARY CONDITIONS AND INITIAL VALUES

The boundary conditions mostly used in our model system are four tidal level components: M2, S2, O1, K1 which are results of the harmonic analysis of tide, the model components, salinity and water temperature, surface wind, air temperature, nutrients, fresh water discharge from river, and intensity of light. They are prescribed based on field measurements at the pertinent observatory. Most of the data are interpolated by the Spline method or linear interpolation. Initial values of components in the new ecosystem model were set on April 1, 2000. The temporal-dependence of these boundary conditions becomes the forcing function of the simulation.

4.3. SIMULATION PERIOD AND TIME STEP

The simulation period of the ecosystem model covers from April 2000 to March 2001 and the time step is set 10 min to the ecosystem model, and 20 sec to the hydrodynamic model. Time series of the output data of the hydrodynamic model, which is used to the ecosystem model, is picked up to adjust the time step of the ecosystem model.

4.4. PARAMETERS

Equations for biochemical reactions in Tables 6 and 7 include biochemical parameters. We investigated a range of values for these parameters that have been observed or used in other models, and set those parameters within that range or at the same order of investigated values. Relevant biochemical parameters we set in this simulation and abstract of the sources of the parameters are listed in Tables 10 and 11. The parameters, either observed data or calculated based on the observed data in the same period and area as those of this simulation, are set within the known ranges. Such parameters are indicated as "OD" at item "Source" in Tables 10 and 11. The parameters whose values were estimated from literature are indicated as "OE" at item "Source". Values were given to the unknown parameters to tune the simulation. The results of the simulation in terms of the concentrations in the components and fluxes of the biochemical reactions that seemed unreasonable in comparison to the observed values, indicated where the parameters needed tuning. Using these processes, we decided the values for the biochemical parameters. The estimation of appropriate molecular diffusion coefficient in sediments "D_S" is carried out by using molecular diffusion in the diluted solution "D_0" taken from Li and Gregory (1974) and by including the effect of tortuosity which is estimated from porosity by Ullman and Aller (1982). In "D_0", we use exponential interpolation to adjust the coefficients to ambient temperature. The formulation of "D_S" is as follows.

$$D_S = \frac{D_0}{\theta^2}, \quad \theta^2 = \phi \cdot F = \phi \cdot \phi^n, \quad n=1.8 \sim 2.0 \qquad (4.4.1)$$

where, θ = tortuosity, $F = \phi^n$ formation factor.

Coefficients concerned with activity of organism, D'_B, D_I, (refer to equation 3.2.6 and 3.2.7) are formulated as a hyperbolic function of components of suspension feeders and deposit feeders as follows and D_B and α are set at zero.

$$D'_B = \frac{(DFB + SFB)}{(DFB + SFB + HF_B)} \cdot D'_{B\,max} \qquad (4.4.2)$$

$$D'_B = \frac{(DFB + SFB)}{(DFB + SFB + HF_I)} \cdot D_{Im\,ax} \qquad (4.4.3)$$

where, DFB = model component of deposit feeders (mass/L^2-sediment), SFB = model component of suspension feeders (mass/L^2-sediment), $D'_{B\,max}$ = maximal coefficient of biodiffusion (interphase mixing) (L^2-sediment/T), $D_{Im\,ax}$ = maximal coefficient of irrigation (L^2-sediment/T), HF_B = half saturation constant for biodiffusion of interphase mixing (mass/L^2-sediment), HF_I = half saturation constant for irrigation (mass/L^2-sediment). If the limiting factors, $SFB + DFB$, has a value HF_B or HF_I, the process proceeds at half the maximal speed; as the components SFB or DFB increases, the limitation becomes weaker (or the value of the limitation function increases).

Table 10. Parameters used in the pelagic system

Name	Unit	Value	Description	Major Source
[Phytoplankton]				
α_{11}, β_{11}	1/hr, 1/degree	0.0246, 0.0693	Maximum growth rate	1, Q_{10}=2
Hf_{n01}	μgN/ml	0.0182	Half saturation constant for nitrogen limitation	2
Hf_{p01}	μgP/ml	0.00093	Half saturation constant for phosphorus limitation	3
I_{opt}	μE/m^2/sec	489.3	Photosynthetic light optimum	4
k_{01}, k_{02}	1/cm	0.0061, 0.0065	Light attenuation independent on plankton and POM	OD
$\gamma_{PP1}, \gamma_{PP1}$	(1/cm)/(μgC/ml)	0.0010, 0.0023	Light attenuation depend on phytoplankton	OD
$\gamma_{ZP1}, \gamma_{ZP1}$	(1/cm)/(μgC/ml)	0.0023, 0.0031	Light attenuation depend on zooplankton	OD
$\gamma_{POC1}, \gamma_{POC1}$	(1/cm)/(μgC/ml)	0.0046, 0.0018	Light attenuation depend on POM	OD
Rgrm01	μgChl-a/μgC	0.0333	Chl-a:C ratio	5
α_{12}, β_{12}	1/hr, 1/degree	0.00125, 0.0693	Rate of respiration	6
α_{13}, β_{13}	1/hr, 1/degree	0.00042, 0.0693	Rate of natural mortality	Tu, Q_{10}=2
[Zooplankton]				
α_{21}, β_{21}	1/hr, 1/degree	0.0075, 0.0693	Maximum ration	7, Q_{10}=2
λ	cm^3/μgC	6.3	Ivlev's constant	8, 9
Π	μgC/cm^3	0	Feeding threshold	10
e	-	0.7	Assimilation efficiency	11, 12
g	-	0.3	Growth efficiency	9
α_{22}, β_{22}	1/hr, 1/degree	0.0021, 0.0693	Rate of mortality	Tu, Q_{10}=2

Table 10 (Continued)

Name	Unit	Value	Description	Major Source
[Detritus, Dissolved organic carbon]				
α_{31}, β_{31}	1/hr, 1/degree	0.01, 0.0693	Maximum mineralization rate of fast-labile POM	13, 14(OE), $Q_{10}=2$
α_{41}, β_{41}	1/hr, 1/degree	0.0001, 0.0693	Maximum mineralization rate of slow-labile POM	15, $Q_{10}=2$
α_{51}, β_{51}	1/hr, 1/degree	0.00001, 0.0693	Maximum mineralization rate of refractory POM	15(OE), $Q_{10}=2$
α_{61}, β_{61}	1/hr, 1/degree	0.001, 0.0693	Maximum mineralization rate of labile DOM	16, $Q_{10}=2$
α_{71}, β_{71}	1/hr, 1/degree	0.0, 0.0693	Maximum mineralization rate of refractory DOM	Tu, $Q_{10}=2$
Kw_{sO2}	mgO_2/l	0.096	Half saturation constant for O_2 limitation in oxic mineralization	17
$Kw^{Sub}_{in\,O2}$	mgO_2/l	0.32	Half saturation constant for O_2 inhibition in suboxic mineralization	17
$Kw^{An}_{in\,O2}$	mgO_2/l	0.16	Half saturation constant for O_2 inhibition in anoxic mineralization	17
$Kw^{Sub}_{s\,NO3}$	mgN/l	1.86	Half saturation constant for NO_3 limitation in suboxic mineralization	17(OE)
$Kw^{An}_{in\,NO3}$	mgN/l	0.5	Half saturation constant for NO_3 inhibition in anoxic mineralization	17(OE)
R_{min03}	-	0.25	Fraction of bio-degradation of fast-labile POM	14
R_{min04}	-	0.25	Fraction of bio-degradation of slow-labile POM	14
R_{min05}	-	0.25	Fraction of bio-degradation of refractory POM	14
[NH_4-N, $NO_{2,3}$-N]				
α_{81}, β_{81}	1/hr, 1/degree	0.003, 0.0693	Rate of nitrification	18(OE), $Q_{10}=2$
Hf_{wno1}	mgO_2/l	0.032	Half saturation constant for O_2 limitation in nitrification	17
A_{wyy}	-	0.5	Denitrification ratio to suboxic mineralization	Tu
[ODU]				
$\alpha_{111}, \beta_{111}$	1/hr, 1/degree	0.1, 0.0693	Rate of oxidation	17(OE), $Q_{10}=2$
Hf_{wouo}	mgO_2/l	0.032	Half saturation constant for O_2 limitation in oxidation	17
A_{sanka}	-	0.3	Fraction of oxidization of anoxic mineralization	Tu
$\alpha_{112}, \beta_{112}$	1/hr, 1/degree	0.0, 0.0693	Rate of authigenic mineralization	Tu

Table 10. (Continued).

Name	Unit	Value	Description	Major Source
A_{wauth}	-	0.0	Fraction of authigenic mineralization	Tu
A_{waut2}	-	0.0	Fraction of authigenic mineralization	Tu
[Eelgrass]				
R_{gwl3}	-	0.6	Mass in water : total mass ratio	OD
α_{131}	1/hr	0.01	Maximum growth rate	29
T_{opt13}	degree	25.0	Photosynthesis temperature optimum	30
A_{sl131}, A_{sl132}	1/degree^2	0.002, 0.0055	Photosynthesis temperature dependency coefficient	30
Hf_{n13}	mgN/l	0.1	Half saturation constant for nitrogen limitation	31
Hf_{p13}	mgP/l	0.08	Half saturation constant for phosphorus limitation	31
I_{opt13}	µE/m^2/sec	80	Photosynthesis light optimum	OD
A_{1306}	-	0.1	Extra-release ratio to photosynthesis	34(OE)
A_{13081}	-	0.2	Rest respiration rate ratio to maximum growth rate	35
A_{13082}	-	0.1	Active respiration ratio to photosynthesis	19(OE)
$\alpha_{132}, \beta_{132}$	1/hr, 1/degree	0.0, 0.0693	Rate of mortality	Tu, $Q_{10}=2$
[Ulva]				
α_{141}	1/hr	0.02	Maximum growth rate	36,37(OE)
T_{opt14}	degree	32.0	Photosynthesis temperature optimum	38
A_{sl141}, A_{sl142}	1/degree^2	0.0016, 0.012	Photosynthesis temperature dependency coefficient	38
Hf_{n14}	mgN/l	0.05	Half saturation constant for nitrogen limitation	Tu
Hf_{p14}	mgP/l	0.03	Half saturation constant for phosphorus limitation	Tu
I_{opt14}	µE/m^2/sec	600.0	Photosynthesis light optimum	37(OE)
A_{1406}	-	0.1	Extra-release ratio to photosynthesis	34(OE)
A_{14081}	-	0.1	Rest respiration rate ratio to maximum growth rate	29,36,39(OE)
A_{14082}	-	0.05	Active respiration ratio to photosynthesis	19(OE)
$\alpha_{142}, \beta_{142}$	1/hr, 1/degree	0.0, 0.0693	Rate of mortality	Tu, $Q_{10}=2$

Table 10. (Continued).

Name	Unit	Value	Description	Major Source
[Epiphite]				
$\alpha_{161}, \beta_{161}$	1/hr, 1/degree	0.0246, 0.0693	Maximum growth rate	1, $Q_{10}=2$
Hf_{n16}	mgN/l	0.0182	Half saturation constant for nitrogen limitation	2
Hf_{p16}	mgP/l	0.00093	Half saturation constant for phosphorus limitation	3
I_{opt16}	$\mu E/m^2/sec$	489.3	Photosynthesis light optimum	4
$\alpha_{162}, \beta_{162}$	1/hr, 1/degree	0.00125, 0.0693	Rate of respiration	6
$\alpha_{163}, \beta_{163}$	1/hr, 1/degree	0.00042, 0.0693	Rate of mortality	Tu, $Q_{10}=2$

Name	Unit	Value	Description	Major Source
[Epifauna]				
$\alpha_{171}, \beta_{171}$	1/hr, 1/degree	0.0075, 0.0693	Maximum ration	7, $Q_{10}=2$
λ_{17}	$cm^3/\mu gC$	6.3	Ivlev's constant	8, 9
Π_{17}	$\mu gC/cm^3$	0.1	Feeding threshold	10
e_{17}	-	0.7	Assimilation efficiency	11, 12
g_{17}	-	0.3	Growth efficiency	9
$\alpha_{172}, \beta_{172}$	1/hr, 1/degree	0.0021, 0.0693	Rate of mortality	Tu, $Q_{10}=2$

Source : 1. Eppley(1972); 2. Eppley et al.(1969); 3. Jørgensen et al.(1991); 4. Ryther(1956); 5. Strickland(1965); 6. Jørgensen (1979); 7. Kremer and Nixon(1975); 8. Zillioux(1970); 9. Suschenya(1970); 10. Smayda(1973); 11. Marshall and Orr(1955a); 12. Marshall and Orr(1955b); 13. Matsunaga(1981); 14. Ishikawa and Nishimura(1983); 15. Ogura(1972); 16. Ogura(1975); 17. Soetaert et al.(1996); 18. Knox et al. (1986); 19. Barreta and Ruardij(1988); 20. Valiela(1984); 21. Kremer and Nixon(1977); 23. Emerson and Hedges(1988); 24. Cammen(1980); 25. Isono et al.(1998); 26. Admiraal et al.(1982); 27. Admiraal et al.(1977); 28. Admiraal et al.(1984); 29. Enriquez et al.(1995); 30. Bulthuis(1987); 31. Udy et al.(1999);32. Dennison and Alberte(1982, 1985, 1986); 33. Dennison(1987); 34. Mukai(1995); 35. Drew(1979); 36. Beach et al.(1995), 37. Coutinho and Zingmark (1993), 38. Einav et al.(1995), 39. Franklin(1994), OD. Observation data of Aichi Prefecture; Tu. Tuning. (OE. Order estimated value referred to the literature).

*Some of the values of these sources were used not directly, but used after analysis to convert the format of model parameters.

Table 11. Parameters used in the benthic system

Name	Unit	Value	Description	Major Source
[Suspension feeder]				
$Wwet_{51}$	gw/ind	0.436	Basal weight	OD
Rwd_{51}	gd/gw	0.019	Dry weight : wet weight ratio	OD
Rcd_{51}	gCdry/gdry	0.328	Carbon : dry weight ratio	OD
$deDO_{51}$	-	0.05	Oxygen saturation ocurring suspension feeder's mortality from oxygen deficiency	Tu
dt	hr	0.1666667	Time step	
dzzSF	cm	7-97	Thickness of bottom layer of pelagic system	
AgrtSF	1/hr	0.002208	Efficient growth rate	25
AbasSF	1/hr	0.0001256	Basal metabolism	19
$Aege_{51}$	-	0.05	Digestivity efficiency	20
$Aexc_{51}$	-	0.9	Activity metabolism	19
β_{512}	1/degree	0.0693	Temperature dependency of mortality	$Q_{10}=2$
Ade_{511}	1/hr	0.00017	Natural mortality rate at 0 degree Celsius	Tu
Ade_{512}	1/hr	0.0017	Rate of mortality from oxygen deficiency	Tu
[Deposit feeder]				
$\alpha_{521}, \beta_{521}$	1/hr, 1/degree	0.0058, 0.0693	Maximum ration	24(OE), $Q_{10}=2$
$Teat_{52}$	degree	10	Maximum saturation temperature about eating	21
$Aivl_{52}$	$cm^3/\mu gC$	0.0005	Ivlev's constant	19(OE)
$Akai_{52}$	$\mu gC/cm^3$	200	Feeding threshold	19(OE)
Hf_{ea52}	$\mu gC/cm^2$	150	Half saturation constant for cannibalism efficiency	19(OE)
$deDO_{52}$	-	0.01	DO saturation occurring deposit feeder's mortality from oxygen deficiency	Tu
Hf_{du52}	$\mu gC/cm^2$	2500.0	Half saturation constant for digestive efficiency	19(OE)
$Andg_{52}$	-	0.4	Minimum undigestive efficiency	20
$Aexc_{52}$	-	0.85	Ratio of excretion to assimilated food	19
$\beta 522$	1/degree	0.0693	Temperature dependency of mortality	$Q10=2$
Ade521	1/hr	0.00025	Natural mortality rate at 0 degree Celsius	Tu
Ade522	1/hr	0.0025	Rate of mortality from oxygen deficiency	Tu

Table 11. (Continued).

Name	Unit	Value	Description	Major Source
[Detritus, Dissolved organic carbon]				
l	cm	0.5	Thickness of 1st layer	
Ko_{sO2}	mgO_2/l	0.096	Half saturation constant for O_2 limitation in oxic mineralization	17
$Ko^{Sub}_{in\,O2}$	mgO_2/l	1.00	Half saturation constant for O_2 inhibition in suboxic mineralization	17(OE)
$Ko^{An}_{in\,O2}$	mgO_2/l	0.16	Half saturation constant for O_2 inhibition in anoxic mineralization	17
$Ko^{Sub}_{s\,NO3}$	mgN/l	0.16	Half saturation constant for NO_3 limitation in suboxic mineralization at oxic layer	17(OE)
$Ko^{An}_{in\,NO3}$	mgN/l	0.32	Half saturation constant for NO_3 inhibition in anoxic mineralization at oxic layer	17(OE)
$Kb^{Sub}_{s\,NO3}$	mgN/l	0.3	Half saturation constant for NO_3 limitation in suboxic mineralization at anoxic layer	17(OE)
$Kb^{An}_{in\,NO3}$	mgN/l	0.2	Half saturation constant for NO_3 inhibition in anoxic mineralization at anoxic layer	17(OE)
$\alpha_{531}, \beta_{531}$	1/hr, 1/degree	0.00050, 0.0693	Maximum mineralizatoin rate of FBP	23, Q_{10}=2
$\alpha_{541}, \beta_{541}$	1/hr, 1/degree	0.000005, 0.0693	Maximum mineralization rate of SBP	23, Q_{10}=2
$\alpha_{551}, \beta_{551}$	1/hr, 1/degree	0.0000001, 0.0693	Maximum mineralization rate of GPM	23(OE), Q_{10}=2
$\alpha_{561}, \beta_{561}$	1/hr, 1/degree	0.001, 0.0693	Maximum mineralization rate of BMM	23, Q_{10}=2
$\alpha_{571}, \beta_{571}$	1/hr, 1/degree	0, 0.0693	Maximum mineralization rate of GMM	Tu, Q_{10}=2
ρ	$g/cm^3 D$	2.34	Average density of total solid	OD
KBMM	ml/g	8.91	Adsorption constant of labile DOM	Tu
KGMM	ml/g	3.00	Adsorption constant of refractory DOM	Tu
Rmin53	-	0.25	Fraction of bio-degradation of FBP	14
Rmin54	-	0.25	Fraction of bio-degradation of SBP	14
Rmin55	-	0.25	Fraction of bio-degradation of GPM	14
[NH_4-N, $NO_{2,3}$-N]				
$\alpha_{581}, \beta_{581}$	1/hr, 1/degree	0.003, 0.0693	Rate of nitrification	18(OE), Q_{10}=2
Hf_{nio1}	mgO_2/l	0.032	Half saturation constant for O_2 limitation in nitrification	17
A_{yyy}	-	0.5	Denitrification ratio to suboxic mineralization	Tu

Table 11. (Continued).

Name	Unit	Value	Description	Major Source
[ODU]				
α_{611}, β_{611}	1/hr, 1/degree	0.1, 0.0693	Rate of oxidation	17(OE), $Q_{10}=2$
Hf_{oduo}	mgO_2/l	0.032	Half saturation constant for O_2 limitation in oxidation	17
A_{sanka}	-	0.3	Fraction of oxidization of anoxic mineralization	Tu
α_{612}, β_{612}	1/hr, 1/degree	0.0001, 0.0693	Rate of authigenic mineralization	Tu
A_{authg}	-	0.5	Fraction of authigenic mineralization	Tu
A_{auth2}	-	0.625	Fraction of authigenic mineralization	Tu
[Benthic algae]				
α_{631}, β_{631}	1/hr, 1/degree	0.031, 0.0693	Maximum growth rate	26(OE), $Q_{10}=2$
Hf_{n63}	mgN/l	0.018	Half saturation constant for nitrogen limitation	2
Hf_{p63}	mgP/l	0.003	Half saturation constant for phosphorus limitation	3
I_{opt63}	$\mu E/m^2/sec$	20	Half saturation constant for light limitation	27
Name	Unit	Value	Description	Major Source
A_{6356}	-	0.122	Extra-release ratio to photosynthesis	OD
A_{63581}	-	0.01	Rest respiration rate ratio to maximum growth rate	28(OE)
A_{63582}	-	0.1	Activity respiration ratio to photosynthesis	19(OE)
CBO	cm	2.2	Half biomass emergence depth	OD
α_{632}, β_{632}	1/hr, 1/degree	0.00019, 0.0693	Rate of natural mortality	19
[Others]				
$D'_{B\,max}$	cm^2/hr	0.016	Maximal coefficient of biodiffusion (interphase mixing)	Tu
$D'_{I\,max}$	cm^2/hr	0.2	Maximal coefficient of irrigation	Tu
HF_B	$\mu gC/cm^2$	1000	Half saturation constant for biodiffusion of interphase mixing	Tu
HF_I	$\mu gC/cm^2$	1000	Half saturation constant for irrigation	Tu

Source : 1. Eppley(1972); 2. Eppley et al.(1969); 3. Jørgensen et al.(1991); 4. Ryther(1956); 5. Strickland(1965); 6. Jørgensen (1979); 7. Kremer and Nixon(1975); 8. Zillioux(1970); 9. Suschenya(1970); 10. Smayda(1973); 11. Marshall and Orr(1955a); 12. Marshall and Orr(1955b); 13. Matsunaga(1981); 14. Ishikawa and Nishimura(1983); 15. Ogura(1972); 16. Ogura(1975); 17. Soetaert et al.(1996); 18. Knox et al.(1986); 24. Cammen(1980); 19. Barreta and Ruardij(1988); 20. Valiela(1984); 21. Kremer and Nixon(1977); 23. Emerson and Hedges(1988); 25. Isono et al.(1998); 26. Admiraal et al.(1982); 27. Admiraal et al.(1977); 28. Admiraal et al.(1984); 29. Enriquez et al.(1995); 30. Bulthuis(1987); 31. Udy et al.(1999);32. Dennison and Alberte(1982, 1985, 1986); 33. Dennison(1987); 34. Mukai(1995); 35. Drew(1979); OD. Observation data of Aichi Prefecture; Tu. Tuning. (OE. Order estimated value referred to the literature).

*Some of the values of these sources were used not directly, but used after analysis to convert the format of model parameters.

4.5. VALIDATION - COMPARISON BETWEEN CALCULATIONS AND OBSERVATIONS

The simulated each model components were compared with the time series of measurements recorded at every monitoring station in Jinno. Here, the comparison between the calculated values and observed values in (i, j)=(3, 4) mesh (refer to Figure 1) is presented in Figure 6 for phytoplankton, POC, DOC, NH_4-N, $NO_{2,3}$-N, PO_4-P and dissolved oxygen (DO) in the pelagic system. Likewise, the comparison in (i, j)=(2, 5), (3, 2), and (4, 5) meshes are presented in Figure 7 for benthic algae, suspension feeders, deposit feeders, POC, NH_4-N, $NO_{2,3}$-N and PO_4-P in the benthic system. Considering the inherent uncertainties in the computational setting of prescribed (forcing) functions and biological parameters, plus the precision of the observations, the model results generally agree with the field observed values. The oscillation of day-time scale (the oscillation of diurnal and tidal cycle) in phytoplankton, NH_4-N and PO_4-P are caused by fluxes of photosynthesis and sea-water transport from the offshore boundary. On the other hand, the oscillation of day-time scale of POC and DOC in the pelagic system are caused only by sea-water transport from offshore boundary. Thus, the amplitude of day-time scale oscillation in POC is smaller than that in phytoplankton. A gap in the time period between light intensity and tidal level (tidal current) also causes oscillation of short-time scale, because this gap influences light intensity in the water that is concerned with depth from surfaces, and photosynthesis rates vary. The calculated values of NH_4-N, PO_4-P in the pelagic system has a tendency of being higher than the observed value at all times. Furthermore, in the case of dissolved oxygen, the calculated values appear to be lower than the observed values. These differences can be explained by the fact that all our comparative field data were collected during daytime. During daytime, photosynthesis of phytoplankton and sea-grass use NH_4-N, PO_4-P and produce dissolved oxygen. Thus, it is reasonable to compare lower calculated value of oscillation of day-time scale to observed value in the case of nutrients (NH$_4$-N, PO$_4$-P) and to compare higher calculated value in the case of dissolved oxygen. The dynamics of NO$_{2,3}$-N in the pelagic system are mainly affected by fluxes of photosynthesis and nitrification.

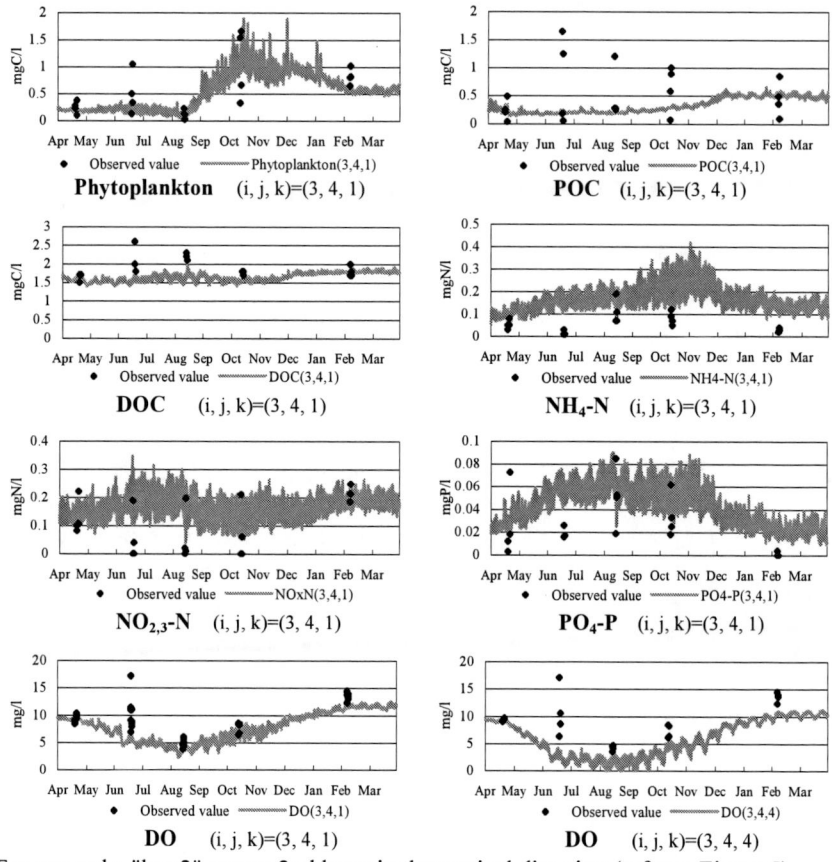

For example, "k = 2" means 2nd layer in the vertical direction (refer to Figure 5).

Figure 6. Comparison between calculation value and observation value in pelagic system (i, j, k) indicates the coordinates of the mesh.

Implementation 51

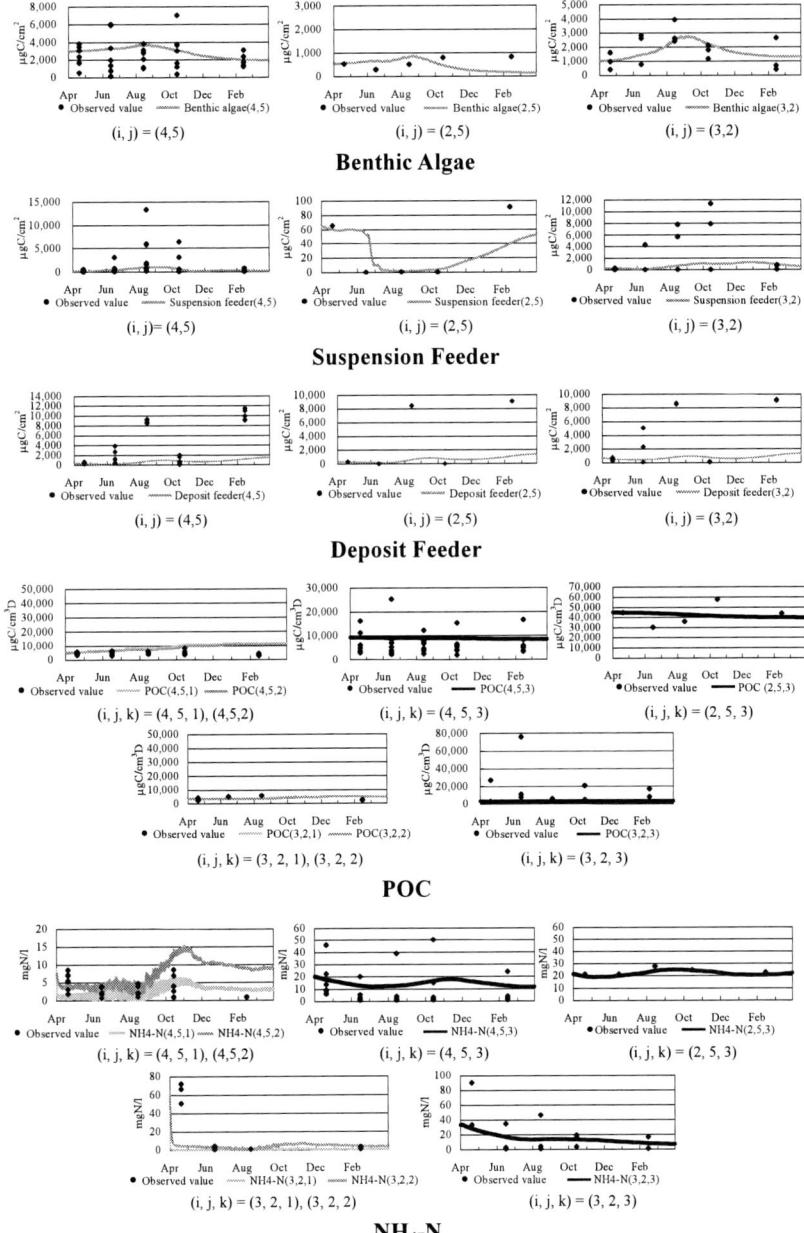

For example, "k = 2" means 2nd layer in the vertical direction (refer to Figure 5).

Figure 7. Comparison between calculation value and observation value in benthic system (i, j, k) indicates the coordinates of the mesh.

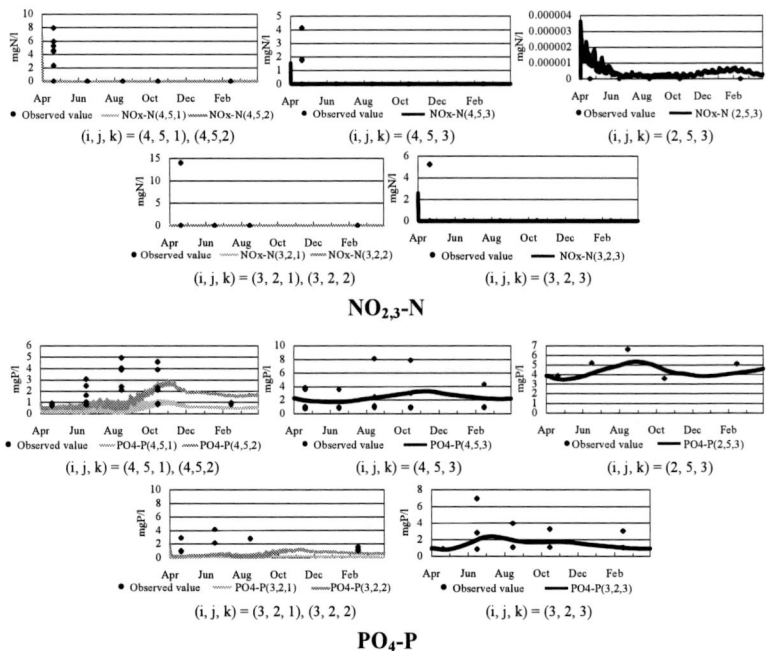

For example, "k=2" means 2nd layer in the vertical direction (refer to Figure 5).

Figure 7 (continued). Comparison between calculation value and observation value in benthic system (i, j, k) indicates the coordinates of the mesh.

The nitrification rate is formulated with the oxygen concentration (given in Tables 6 and 7) and during the daytime, this flux becomes greater because of increasing oxygen caused by photosynthesis. On the other hand, photosynthesis uses nitrate ($NO_{2,3}$-N) during daytime. The concentration of $NO_{2,3}$-N is determined from the balance of these increasing and decreasing fluxes. The seasonal trend of benthic algae is mainly under the control of light intensity and water-temperature. These control factors become weaker/lower in winter than in summer. Therefore, biological activity becomes lower in winter than in summer. The rapid decreasing of suspension feeders at (i, j) = (2,5) in summer comes from mortality caused by oxygen depletion in the bottom layer. The impact of oxygen depletion on the mortality of deposit feeders is less conspicuous than on suspension feeders because the tolerance of deposit feeders to oxygen depletion is generally much greater than that of suspension feeders so we set biochemical parameters as such. The oscillation of day-time scale of NH_4-N and PO_4-P of the surface in pore water are caused by photosynthesis of benthic algae.

Chapter 5

EVALUATION OF PART 1 - LIKELIHOOD OF SEAGRASS GROWTH ON THE ARTIFICIAL SHALLOWS

As explained in the section *"2.2. Evaluation Program"*, the evaluation of Part 1 was performed by using the TRÄUMEREI model applied to the Jinno area shown in Figure 1. The details of the model construction, implementation (initial values, boundary conditions, and biological parameters) and the model validation were already described in Sections *"3. Ecosystem Modeling"* and *"4. Implementation"*. In this section, the results for scenarios in Table 1 derived from the TRÄUMEREI model analysis are discussed.

5.1. ECOLOGICAL MECHANISM IN THE NATURAL SEAGRASS BEDS

One of the methods to reveal the aspects of seagrass beds ecosystem is to compare its ecological mechanism with the mechanisms of different ecosystems from seagrass such as shallow water area without seagrass or hypoxic area. The Jinno area where the TRÄUMEREI model applied is comprised of seagrass beds, shallow waters without seagrass near the mouth of the river, and an oxygen depleted offshore area. As shown in figure 1, Seagrass beds are located at the coordinates, (i, j) = (4, 4), (4, 5). The submerged/emerged area is located at (i, j) = (4, 3). Freshwater flow from the Toyokawa River enters the Bay from (i, j) = (2, 5).

Here, to investigate differences in the ecological mechanisms between areas of seagrass beds (area 1), shallow waters without seagrass near the mouth of the river (area 2), and an oxygen depleted offshore area (area 3), we analyzed how nutrients are cycled in each area. In the analysis, we assumed that the averaged results of each set of grids (meshes), (i, j) = (4, 4), (4, 5), (i, j) = (4, 2), (4, 3) and (i, j) = (2, 3), (2, 4) would be representative of areas 1, 2 and 3 as shown in Figure. 1. The nutrient cycling was analyzed in each of the three sets.

5.1.1. Biological and Physical Processes in Nutrient Cycling

The details of nutrient cycling of these areas, where the cycling is the time-averaged value in June, are presented in Figures 8, 9 and 10. The reason of evaluation period, June, is that the biomass of seagrasses is highest in June and is suitable to reveal the characteristics/functions of seagrass beds. In addition, a simplified nutrient cycling in June is presented in Figure 11. Here, model components and fluxes were arranged by emphasizing forms of nitrogen in Figures 8 to 11. The nitrogen cycles indicated in Figures 8 to 11 are coupled with carbon, phosphorus, and oxygen cycles in the model calculation. The detail of association between new components, <A>,<B>,<C>,<D>,<E> in Figure 11 and model components in Figure 2 are given in Table 12. The association between fluxes concerned with new components in Figure 11 and detail fluxes in Figures. 8, 9 and 10 are also given in Table 13. From these figures, we tried to understand the characteristics of the ecological mechanism in each of the three areas. The results are presented in the following.

(1) Fluxes at System Boundary:
In area 1 (seagrass bed area), organic nitrogen (total fluxes of <A> and <B> at the system boundary in Figure 11) is transported out of system and inorganic nitrogen (<C> in Figure 11) is brought in, though the reverse situation occurred at areas 2 and 3. These difference between area 1 and area 2 or area 3 can be explained by high organic production in the seagrass beds (area 1), especially the production associated with epiphytes and epifauna.

(2) Fluxes between <A> and <D> in Figure 11:
Many suspension feeders inhabit areas 1 and 2. Thus, the biological flux from <A> to <D> in areas 1 and 2 is large compared to that in area 3. On the

other hand, with regard to the physical flux, the flux from <D> to <A> in area 2 is larger than in the other two areas because of the high re-suspension rate, which is caused by the kinetic energy of fresh-water inflow from the river.

(3) Turnover Rate of <D> in Figure 11:

Each fluxes concerned with <D> (input flux to <D> or output flux from <D>) at areas 1 and 2 are mostly larger than those of area 3. However, the mass of <D> at area 3 is larger than at areas 1 and 2. This condition means that turnover rate of <D> in areas 1 and 2 are higher than area 3. This phenomenon can be explained that in area 1 and area 2, there are many suspension feeder, and many quantity of their feeding flux lead organic matter, including both living and non-living which are food of suspension feeder, to sediment quickly. Thus, most of the constituents of organic matter carried to the sediment are composed of fast labile organic matter, which is part of every living cell (cell of phytoplankton, zoo-plankton, sea-grass, seaweed, suspension feeders and deposit feeders, epiphytes, epifauna) and of non-living carbons (detritus and DOM). This is because these organic matters (fast labile organic matter) are carried to the sediment before they are mineralized in water. Hence, organic matters at sediment are mineralized comparatively quickly than it is in area 3.

(4) Mineralization Pathway and De-nitrification in the Benthic System:

The ratio (appears as a percentage in Table 14) of mineralization pathway in the benthic system and the concentration of dissolved oxygen in each of the three study areas are given in Table 14 and Table 15. In area 1, the concentration of dissolved oxygen in the bottom water is comparatively high because of more photosynthesis than other areas due to epiphytes and benthic algae (In areas 2 and 3, there are no epiphytes. Photosynthesis of benthic algae is weaker in area 3 compared to areas 1 and 2, because light intensity at the sea bottom is low in area 3). Therefore, the ratio of oxic mineralization per all forms of mineralization in the benthic system becomes high. On the other hand, in area 3, a low concentration of oxygen in the bottom water leads to the high ratio of anoxic mineralization. In area 2, the ratio of suboxic mineralization is high compared to the other areas. This situation can be explained by the concentration balance of oxygen and nitrogen (refer to the formulation in Table 7). The high ratio of suboxic mineralization leads to the high de-nitrification rate in area 2.

The differences in thickness of the diffusive boundary layer (DBL), which was tuned according to the quantity of the horizontal velocity, has an influence on the $NO_{2,3}$ diffusion transport from the pelagic to the benthic system. This

56   Akio Sohma

diffusion is also one of the causes of high suboxic mineralization of the benthic system in area 2. DBL in area 2 is set to be thinner than the other areas due to its closer proximity to the river.

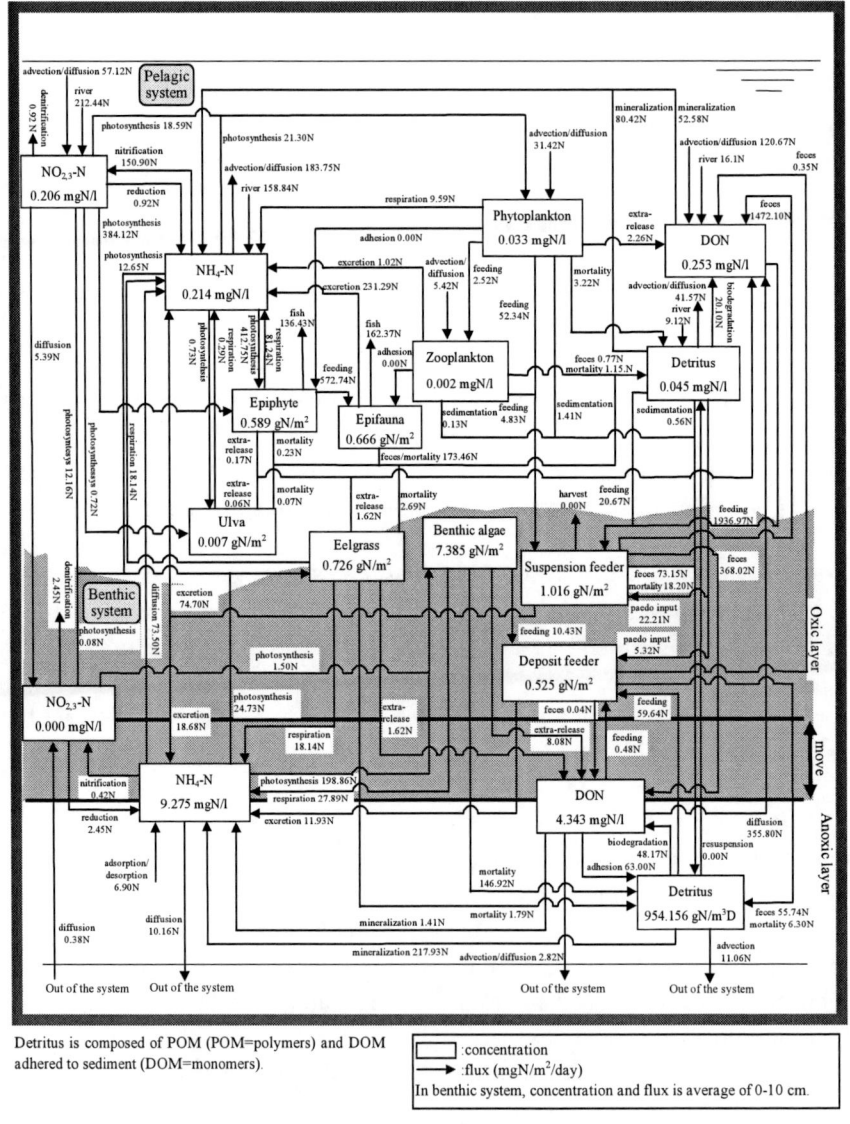

Figure 8. Nitrogen cycle in area 1 (sea-grass (eelgrass) beds) in June.

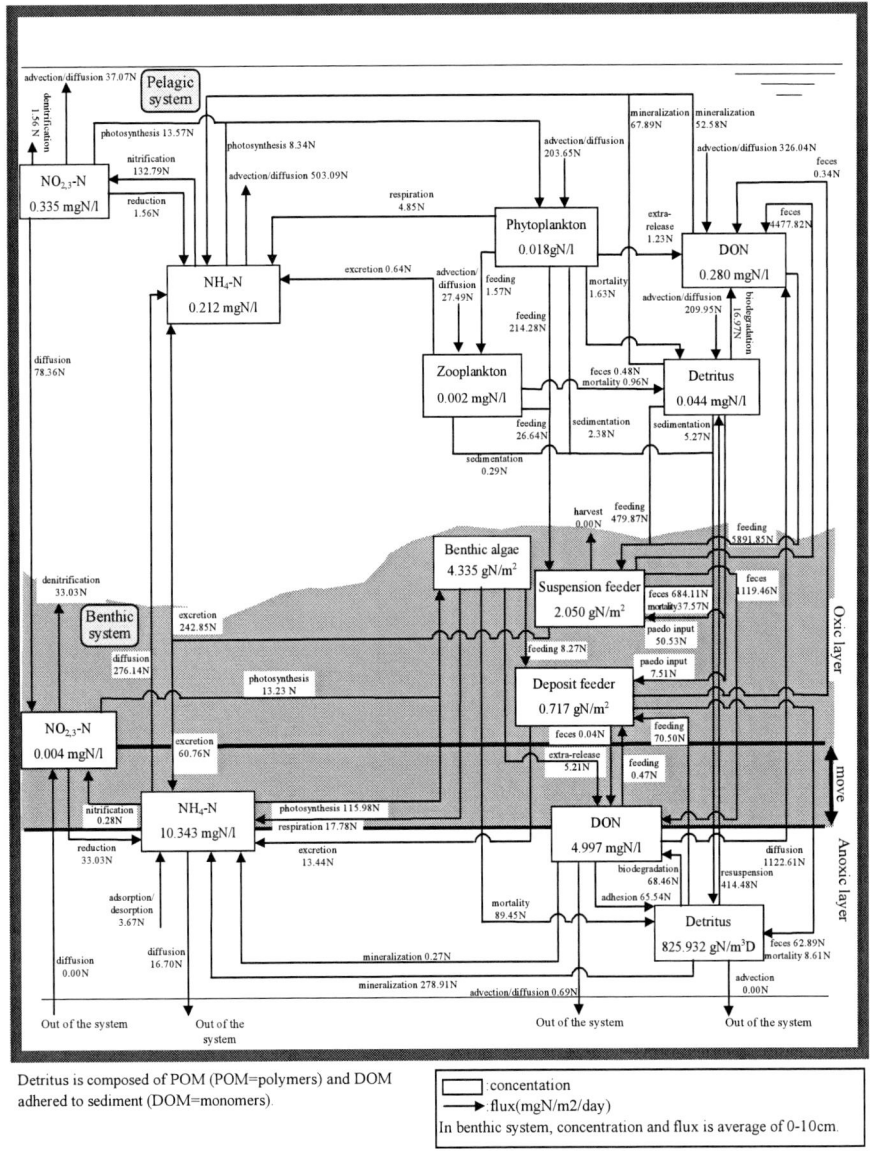

Figure. 9. Nitrogen cycle in area 2 (shallow waters without sea-grass) in June.

(5) Overall Image of Vertical Cycle for Each Area:

An overall image of nitrogen cycle for each area is described in the following:

(a) In sea-grass beds (area 1), suspension feeders fix fresh organic matter at the bottom of sea-grass beds. Most of the fixed organic matter has a fast mineralization rate and mineralizes quickly. As a result, they supply the nutrients for production of sea-grass, benthic algae and epiphytes. In addition, transparency of sea-water becomes clear and penetration depth of light becomes deeper due to feeding of suspension feeders. For these reasons, we suggest that the biological processes associated with suspension feeders are important for the production of sea-grass.

(b) In shallow waters without sea-grass near the mouth of a river (area 2), there are many suspension feeders and many feeding of fresh organic matter as in area 1. Thus, fresh organic matter in the pelagic system is transported to the benthic system. However, because of high rates of re-suspension caused by the kinetic energy of water inflow, the fresh organic matter is re-suspended, and does not settle in the benthic system for a long time. However, some re-suspended fresh organic matter is consumed by suspension feeders and is transported to the benthic system again.

(c) In the oxygen depleted offshore area (area 3), there are few macro benthos as they cannot survive without oxygen so the turnover rate of several components is low.

From above consideration, sea-grass beds have a remarkable vertical cycle of nitrogen by biological processes due to suspension feeders and photosynthesis by sea-grass and epiphytes. Furthermore, shallow waters without sea-grass near the mouth of the river also has vertical cycle of nitrogen through physical and biological processes associated with re-suspension and feeding by suspension feeders. However the oxygen depleted offshore area has no remarkable vertical cycle of nitrogen.

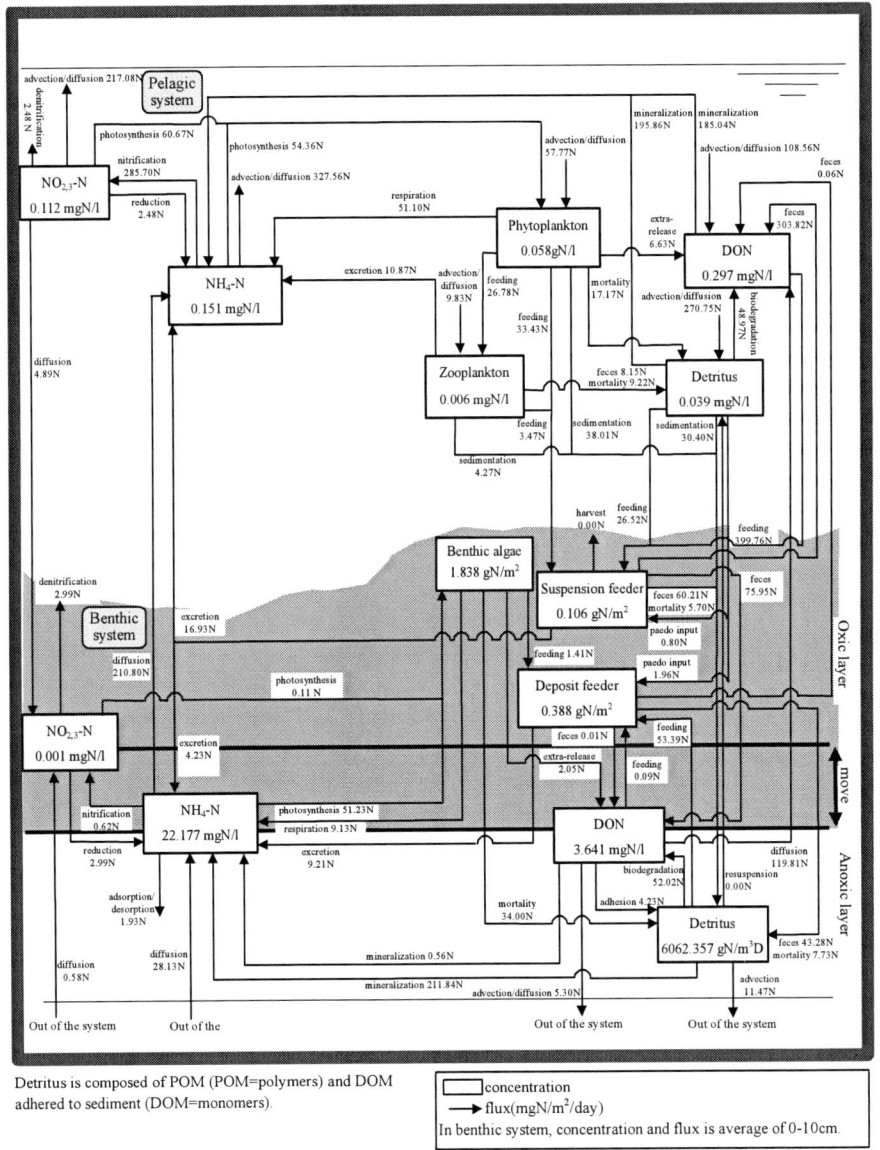

Figure 10. Nitrogen cycle in area 3 (oxygen depleted offshore area) in June.

Table 12. Classification of component in simplified cycle

	area 1 Sea-grass beds	area 2 Shallow waters without see-grass	area 3 Oxygen depleted offshore area
A : Pelagic organic nitrogen (suspended)	Phytoplankton Zooplankton Detritus DON	Phytoplankton Zooplankton Detritus DON	Phytoplankton Zooplankton Detritus DON
B : Pelagic organic nitrogen (non-suspended)	Eelgrass Ulva Laver Epiphyte Epifauna		
C : Pelagic inorganic nitrogen	NH_4-N $NO_{2,3}$-N	NH_4-N $NO_{2,3}$-N	NH_4-N $NO_{2,3}$-N
D : Benthic organic nitrogen	Benthic algae Suspension feeder Deposit feeder Detritus DON	Benthic algae Suspension feeder Deposit feeder Detritus DON	Benthic algae Suspension feeder Deposit feeder Detritus DON
E : Benthic inorganic nitrogen	NH_4-N $NO_{2,3}$-N	NH_4-N $NO_{2,3}$-N	NH_4-N $NO_{2,3}$-N

Table 13. Classification of flux in simplified cycle

source/ destination	Biochemical flux	Physical flux
A→B	—	•adhesion of phytoplankton and zooplankton
A→C	•respiration of phytoplankton •excretion of zooplankton •mineralization of detritus and DON	—
A→D	•phytoplankton, zooplankton, detritus, DON feeded by suspension feeder	•sedimentation of pytoplankton, zooplankton and detritus •diffusion of DON at sediment-water interface (if pelagic→benthic) •paedo input
A→E	—	—
A↔Out of the system	•phytoplankton feeded by fish •zooplanton feeded by fish •detritus feeded by fish	•advection/diffusion of phytoplankton, zooplankton, detritus, DON •river load of detritus, DON
source/ destination	Biochemical flux	Physical flux

Table 13. (Continued)

B→A	•extra-release of eelgrass, ulva, laver and epiphyte •mortality of eelgrass, ulva, laver, epiphyte and epifauna •feces of epifauna	—
B→C	•respiration of eelgrass, ulva, laver and epiphyte •excretion of epifauna	—
B→D	•mortality of eelgrass	—
B→E	•respiration of eelgrass	—
B↔Out of the system	•epiphyte feeded by fish •epifauna feeded by fish	•harvest of ulva •advection of epiphyte and epifauna
C→A	•NH_4-N, $NO_{2,3}$-N used by photosynthesis of phytoplankton	—
C→B	•NH_4-N, $NO_{2,3}$-N used by photosynthesis of eelgrass, ulva, laver and epiphyte	—
C→D	—	—
C→E	—	•advection/diffusion of NH_4-N, $NO_{2,3}$-N at sediment-water interface(if pelagic→benthic)
C↔Out of the system	•denitrification	•advection/diffusion of NH_4-N, $NO_{2,3}$-N •river load of NH_4-N, $NO_{2,3}$-N
D→A	•feces of suspension feeder and deposit feeder (DON)	•resuspension of detritus and benthic algae •diffusion of DON at sediment-water interface(if benthic→pelagic)
D→B	—	—
D→C	•excretion of suspension feeder	—
D→E	•respiration of benthic algae •excretion of suspension feeder and deposit feeder •mineralization of detritus and DON	—
D↔Out of the system	•suspension feeder, deposit feeder and benthic algae feeded by fish	•harvest of suspension feeder •bury of detritus •advection/diffusion of DON
E→A	—	—
E→B	•NH_4-N, $NO_{2,3}$-N used by photosynthesis of eelgrass	—
E→C	—	advection/diffusion of NH_4-N, $NO_{2,3}$-N at sediment-water interface (if benthic→pelagic)
E→D	•NH_4-N, $NO_{2,3}$-N used by photosynthesis of benthic algae	—
E↔Out of the system	•denitrification	•advection/diffusion of NH4-N, NO2,3-N

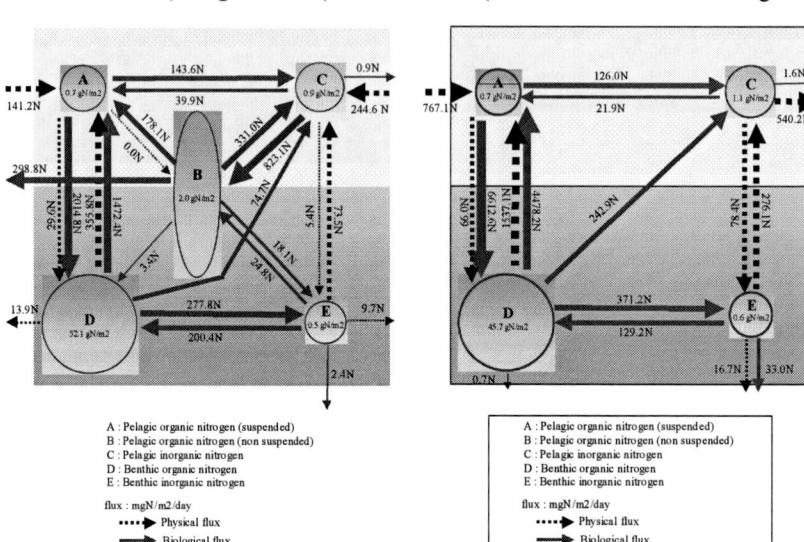
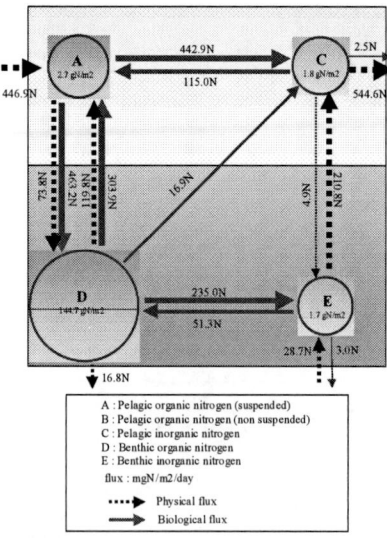

Figure 11. Simplified nitrogen cycle in June.

Table 14. The ratio of mineralization pathway in the benthic system in June

	oxic mineralization	suboxic mineralization	anoxic mineralization
Area 1 (sea-grass beds)	18.9 %	0.8 %	80.3 %
Area 2 (shallow waters without sea-grass)	13.8 %	8.5 %	77.7 %
Area 3 (oxygen depleted offshore area)	4.4 %	1.0 %	94.6 %

Table 15. The oxygen concentration in June

	bottom water value	vertical averaged value
Area 1 (sea-grass beds)	5.8 mgO$_2$/l	6.6 mgO$_2$/l
Area 2 (shallow waters without -grass)	4.5 mgO$_2$/l	4.9 mgO$_2$/l
Area 3 (oxygen depleted offshore area)	0.6 mgO$_2$/l	4.2 mgO$_2$/l

Table 16 (1). Turnover rate of biological processes in June

	Area 1 (Sea-grass beds)	Area 2 (Shallow waters without sea-grass)	Area 3 (Oxygen depleted offshore area)
A: Pelagic organic nitrogen (suspended)	2.67	8.45	0.24
B: Pelagic organic nitrogen (non suspended)	0.42	-	-
C: Pelagic inorganic nitrogen	0.78	0.19	0.16
D: Benthic organic nitrogen	0.04	0.13	0.00
E: Benthic inorganic nitrogen	0.51	0.48	0.09

Units of 1/day. For each components in this table, shading represents the area where the turnover rate is the maximum for the three areas.

Table 16 (2). Turnover rate of physical process in June

	Area 1 (Sea-grass beds)	Area 2 (Shallow waters without sea-grass)	Area 3 (Oxygen depleted offshore area)
A: Pelagic organic nitrogen (suspended)	0.37	1.78	0.12
B: Pelagic organic nitrogen (non suspended)	0.00	-	-
C: Pelagic inorganic nitrogen	0.18	0.42	0.21
D: Benthic organic nitrogen	0.00	0.02	0.00
E: Benthic inorganic nitrogen	0.09	0.34	0.07

Units of 1/day. For each components in this table, shading represents the area where the turnover rate is the maximum for the three areas.

5.1.2. Turn over Rate of Nutrient Cycling

To describe the driving forces of nutrient cycling quantitatively, the turnover rate of biological processes and physical processes from the model outputs was calculated. The turnover rate was defined as follows (Sohma et al., 2004):

Turnover rate of biological processes (day^{-1})

$$= \frac{0.5 \times \left\{ \binom{\text{Total of input biological fluxes}}{\text{to the component (mgN m}^{-2}\text{ day}^{-1})} + \binom{\text{Total of output biological fluxes}}{\text{from the component (mgN m}^{-2}\text{ day}^{-1})} \right\}}{\left(\text{Mass of the component (mgN m}^{-2})\right)}$$

(5.1.1)

Turnover rate of physical processes (day^{-1})

$$= \frac{0.5 \times \left\{ \binom{\text{Total of input physical fluxes}}{\text{to the compotnent (mgN m}^{-2}\text{ day}^{-1})} + \binom{\text{Total of output physical fluxes}}{\text{from the compotnent (mgN m}^{-2}\text{ day}^{-1})} \right\}}{\left(\text{Mass of the compoment (mgN m}^{-2})\right)}$$

(5.1.2)

The values of the turnover rate in each area in June are given in Table 16. From the table, the characteristics of nutrient cycling in each area are as follows:

(a) The turnover rate of biological and physical processes in shallows with seagrass beds (area 1) and in shallows without seagrass beds near the mouth of a river (area 2) is generally higher than in the oxygen depleted offshore area (area 3). There is one exception where the turnover rate of a physical process (C: pelagic inorganic nitrogen) is higher in area 3 than area 1 (Table 16).

(b) The turnover rate of physical processes (mainly caused by re-suspension) in area 2 is much greater when compared to areas 1 and 3, while the turnover of inorganic matter by biological processes in area 1 is high compared to areas 2 and 3.

(c) The high turnover rate of biological processes for inorganic matter in area 1 results from the high rate of production by living organisms (excretion) and the high rate of consumption (photosynthesis of seagrasses, epiphytes, and benthic algae). In contrast, in area 2, the turnover rate of biological processes for inorganic matter is lower than in area 1 due to the absence of vegetation in this area (no consumption by photosynthesis). In addition, the turnover rate of physical processes is higher than in area 1 reflecting the high rates of physical exchange (a) between the pelagic and the benthic ecosystems due to vertical mixing or (b) between the region of area 2 and out of the region due to horizontal advection or diffusion.

(d) In area 1, the turnover rates for the biological processes for every component (A-E) are higher than those of the physical processes (Table 16). Similarly, in area 2, the turnover rates of the biological processes for A, D, and E are higher than those of the physical processes. These results show that in seagrass beds and in shallows without seagrass beds, biological processes are a more important driving force of nutrient cycling than physical processes. The turnover rates for physical processes are greater in area 2 than in areas 1 and 3 but generally lower than the turnover rates for the biological processes

5.2. EFFECTIVENESS OF SUSPENSION FEEDERS TO SEAGRASS GROWTH

Section 5.1 demonstrates the important role of biological processes to the nutrient cycle in seagrass beds (high turnover rate of biological processes), and represented that the high turnover rate of biological processes is due to the metabolism of suspension feeders. That is, biodeposits from filter feeding benthic fauna (suspension feeders) contributed a high proportion of the total suspended load in the Jinno area. Such biologically mediated sedimentation has the capacity to exceed passive physical processes in the deposition of fresh sediments in the Jinno area.

In order to investigate the effect of the biological processes of the suspension feeders on the metabolism of seagrass, we simulated the scenario without suspension feeders in the Jinno area (scenario 1-2 in Table 1). In this calculation, the biomass of the suspension feeders was set at zero. The other calculation conditions are the same as for scenario 1-1.

A comparison of the average rate of photosynthesis of seagrasses, light intensity and a number of other model components between scenario 1-1 (existing conditions in the Jinno area) and scenario 1-2 (absence of suspension feeders in the Jinno area) in June is shown in Table 17. The increase in phytoplankton and detritus concentrations is clearly due to the absence of suspension feeders. This increase leads to high light attenuation and a decrease in the amount of light reaching the sea floor – an important factor in the rate of photosynthesis of seagrasses (Kawasaki et al., 1990). Concentrations of nutrients such as NH_4-N, $NO_{2,3}$-N, and PO_4-P in scenario 1-2 are low compared to the existing conditions (scenario 1-1). This decrease in nutrient concentrations is due to the fact that there is no excretion from suspension feeders. The difference in the photosynthetic rate of eelgrass (*Zostera marina*) between scenarios 1-1 and 1-2 is only 3.5%. However, photosynthetic flux is formulated as a first order reaction as follows:

$$\frac{dC}{dt} = aC \tag{5.2.1}$$

where, C = Biomass of seagrass (Mass), t = time (T), a = the specific growth rate: photosynthetic rate (T^{-1}). Therefore, as time goes by, the differences in the biomass of eelgrass (*Zostera marina*) become greater than the differences in the specific growth rate (photosynthetic rate).

Table 17. The effect of suspension feeders on the seagrass beds (averaged value in June)

Concentration/fluxes of model components	Unit	With suspension feeders	Without suspension feeders	Effect of suspension feeders
Photosynthesis rate of eelgrass	$\mu gC\ cm^{-2}\ day^{-1}$	33.7	32.5	increase (3.5%)
Light attenuation coefficient	m^{-1}	0.73	0.79	decrease (8.7%)
Nutrient limitation of eelgrass photosynthesis	-	0.940	0.937	increase (0.3%)
Phytoplankton	$mgC\ l^{-1}$	0.143	0.310	decrease (116.6%)
Detritus in pelagic system	$mgC\ l^{-1}$	0.228	0.320	decrease (40.5%)
NH_4-N in pelagic system	$mgN\ l^{-1}$	0.195	0.146	increase (24.9%)
$NO_{2,3}$-N in pelagic system	$mgN\ l^{-1}$	0.160	0.132	increase (17.2%)
PO_4-P in pelagic system	$mgP\ l^{-1}$	0.0584	0.0480	increase (17.8%)
Dissolved NH_4-N in benthic system	$mgN\ l^{-1}$	10.99	10.53	increase (4.2%)
Dissolved PO_4-P in benthic system	$mgP\ l^{-1}$	1.505	1.419	increase (5.7%)

Concentration of each component is averaged vertically at the seagrass beds, (i, j) = (4, 5) in Figure. 1.

From this analysis, the effect of the suspension feeders is illustrated as follows:

(a) Feeding by suspension feeders increases water transparency and improves light conditions for the photosynthesis of seagrass beds. In the pelagic ecosystem, the effect of supplying organic matter through the feces of suspension feeders is smaller than the effect of removing organic matter by feeding.

(b) Excretion by suspension feeders supplies nutrients both to sediments and to the water column, which are necessary for the photosynthesis of seagrasses and epiphytes.

5.3. LIKELIHOOD OF SEAGRASS GROWTH ON THE ARTIFICIAL SHALLOWS IN THE MITO AREA

In scenario 1-3 the TRÄUMEREI model was used to evaluate the suitability of artificial shallows in Mito area for the growth of seagrass beds. The model described the Mito area (Figure. 1) as one grid in the horizontal direction and the grids in the vertical direction were set the same as for scenarios 1-1 and 1-2 in the Jinno area. Other conditions in the model were set as the existing condition observed in the Mito area. The calculation period was the same as in the simulation for the Jinno area. In this calculation, the following three assumptions were used.

(a) *Assumption 1:* The depth of the created artificial shallows was assumed as the datum line minus 0.8 m (D.L. -0.8 m). The datum line is the atmospheric-ocean boundary line at mean tidal level. The set value of D.L. -0.8m was deemed to adequately satisfy the following two conditions and was selected based on the field data collected in the Mito area in 2000. The two conditions are: (a) the depth of the artificial shallows must be higher than the level where the amount of living matter in the benthos does not decrease due to hypoxia and (b) the depth must be higher than the level where enough light reaches the seafloor for seagrasses to photosynthesize (3 E m-2 day-1 of the average light concentration throughout the year). The value of 3 E m-2 day-1 was calculated by Kawasaki et al. (1990).

(b) *Assumption 2:* The initial biomass of benthic fauna and the biomass of the seagrass beds, epiphytes and epifauna were set at the observed value in the Miya area (located 10 km west of the Mito area) (Figure. 1). In the Miya area seagrass beds are currently growing on existing artificial shallows. The biomass of seagrasses in the Miya area is greater than the biomass of the natural seagrass beds in the Jinno area. This assumption will allow the evaluation of the successful cultivation of seagrass beds on the artificial shallows in the Mito area.

(c) *Assumption 3:* The initial concentrations of the detritus in sediments and the nutrients in pore waters were set at the observed value in the Mito area.

The comparison of the results of scenario 1-1 and this scenario (1-3) in June 2000 is shown in Table 18. The results of the calculation for artificial shallows without seagrass in the Mito area are also shown in Table 18. The difference in the assumptions for this calculation compared to scenario 1-3 is that the biomass of seagrass, epiphytes and epifauna were set at zero. The photosynthetic rate (specific growth rate) of eelgrass (*Zostera marina*) in the artificial shallows in the Mito area is greater than for eelgrass in the Jinno area under the assumptions of scenario 1-3 (refer to items A and C in Table 18). This difference in the rate of photosynthesis between the Mito area and the Jinno area is due to the light and benthic nutrient levels (Table 18). The lower light attenuation coefficient and the higher concentrations of benthic nutrients in the artificial shallows with seagrass beds in the Mito area (Table 18) are due to a combination of factors: (a) the differences in water quality between the Mito area and the Jinno area due to geological location; and (b) that the biomass of the benthic living matter (suspension and deposit feeders) is higher in the artificial shallows in the Mito area compared to the natural seagrass beds in the Jinno area (Table 18). The effect of the higher biomass of benthic fauna in the artificial shallows of the Mito area leads to higher rates feeding by suspension feeders. This higher feeding rate removes phytoplankton and detritus more effectively from seawater and transports organic matter to the benthic ecosystem. Moreover, excretion by benthic fauna supplies more nutrients to the benthic ecosystem. However, the concentration of phytoplankton in the Mito area is higher than that in the Jinno area, and the concentrations of nutrients in the pelagic ecosystem in the Mito area is lower than in the Jinno area. These results can be attributed to the geological location of the sites. The Mito area generally supports higher concentrations of phytoplankton and lower concentrations of pelagic nutrients area compared to the Jinno area.

The results of the simulation for scenario 1-3 suggest that if the conditions assumed for this scenario and the stability of the artificial shallows could be achieved, the likelihood of creating artificial shallows on which seagrass beds can be cultivated is high. However, some artificial measures such as seagrass transplantation might be required to cultivate the seagrass beds on the artificial shallows. This model output does not evaluate the natural or autonomous growth of seagrass from the condition of no seagrass but evaluates the photosynthetic rate under the assumption that seagrass is present.

Table 18. Comparison of the seagrass beds in the Jinno and Mito areas (June)

Concentration/fluxes of model components	Unit	A. Jinno area Seagrass beds	B. Mito area Artificial shallows without seagrass, epiphytes and epifauna (D.L. -0.8m)	C. Mito area Artificial shallows with seagrass, epiphytes and epifauna (D.L. -0.8m)	Comparison ((C-A)/A*100)
Photosynthesis rate of eelgrass	day^{-1}	0.034	-	0.036	5.6%
Light attenuation coefficient	m^{-1}	0.73	0.687	0.692	-4.9%
Phytoplankton	mgC l^{-1}	0.132	0.149	0.146	11.2%
Detritus in pelagic system	mgC l^{-1}	0.237	0.097	0.116	-50.9%
NH_4-N in pelagic system	mgN l^{-1}	0.215	0.197	0.137	-36.5%
$NO_{2,3}$-N in pelagic system	mgN l^{-1}	0.207	0.099	0.032	-84.5%
PO_4-P in pelagic system	mgP l^{-1}	0.067	0.043	0.027	-59.6%
Dissolved NH_4-N in benthic system	mgN l^{-1}	9.28	10.51	11.38	22.7%
Dissolved PO_4-P in benthic system	mgP l^{-1}	1.30	2.95	3.03	133.6%
Suspension feeder	μgC cm^{-2}	530.2	4896.0	4920.8	828.0%
Deposit feeder	μgC cm^{-2}	309.6	977.3	977.4	215.7%
Eelgrass	μgC cm^{-2}	851.9	-	4849.5	396.2%

Concentration of each component is averaged vertically. Each value in the Jinno area is an averaged value at (i, j) = (4, 4) and (4, 5). Depth of the created artificial shallows set to D.L. -0.8 m.

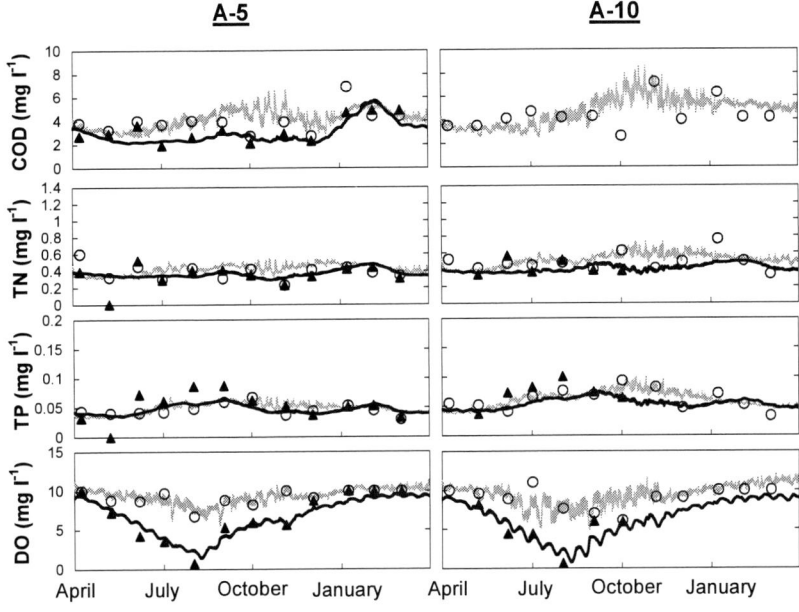

○ **Obs. (Upper layer)** ▲ **Obs. (Bottom layer)** ······ **Calc. (Upper layer)** —— **Calc. (Bottom layer)**

Figure 12. Comparison of the temporal distribution of COD, TN, TP and DO concentrations between model outputs and observed values at station A-5 and A-10 (refer to Figure 1(. Observation depths are 0.5 m for upper layer (○),and seafloor + 0.5 m for bottom layer (▲). Water depths at A-5 and A-10 are 12 m and 9 m, respectively. There is no data for bottom layer of COD at A-10.

Chapter 6

EVALUATION OF PART 2 - EFFECT OF RECLAIMING/CREATING THE SEAGRASS BEDS ON ATSUMI BAY

The calculation conditions and results for scenarios in Table 2 are discussed in this section.

6.1. MODEL VALIDATION AND SPATIAL DISTRIBUTION OF WATER QUALITY

In Part 2, the NCME model was applied to Atsumi Bay excluding the Jinno and Mito areas. The TRÄUMEREI model was then applied to the Jinno and Mito areas where natural or simulated seagrass beds are present. Here, the output from the TRÄUMEREI model was used as the boundary condition for the NCME model at the boundary of the Jinno and Mito areas. The construction of the NCME model and the calculation conditions (initial conditions, boundary conditions at the mouth of the bay, and the value of the biological parameters) were described in a previous paper (Sohma et al., 2001). The difference in the simulation is the scale and the spatial resolution at which it is applied. In the earlier study, the NCME model was applied to Mikawa Bay, which includes Atsumi Bay, divided into eight boxes. However in this evaluation, the model was applied only to Atsumi Bay and the calculation was implemented with finer grids in the horizontal (500 m to 2 km meshes) and vertical planes as shown in Figure. 1.

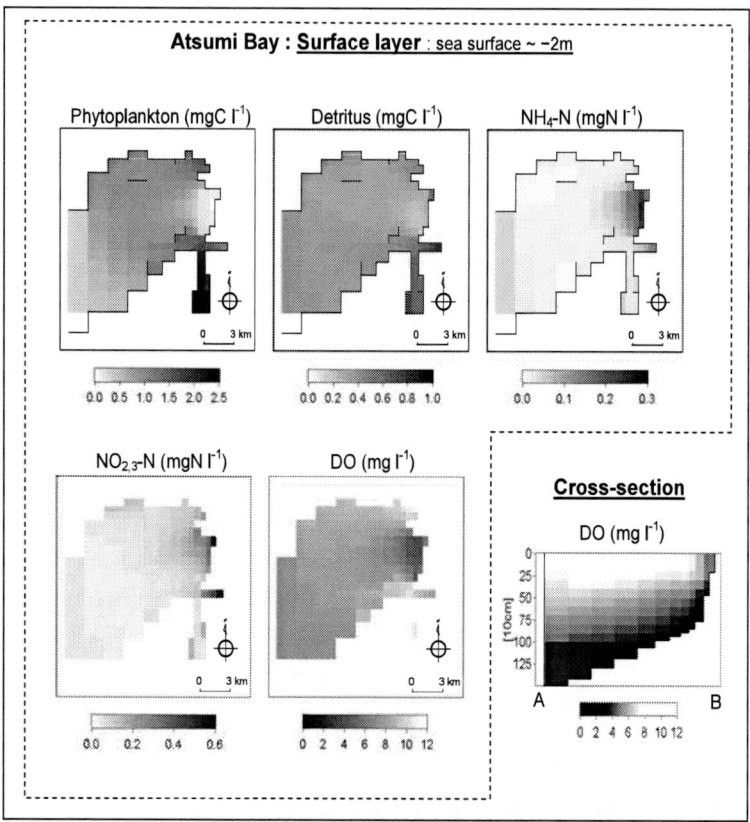

Figure 13. Spatial distribution of model components in the existing conditions in Atsumi Bay during summer (averaged value of July, August and September). The location of the cross section of DO is indicated by a dashed line A-B in Figure 1. The right side of the figure in the cross-section corresponds to point B.

The calculated values of COD (chemical oxygen demand), TN (total nitrogen), TP (total phosphorus), and DO (dissolved oxygen) in scenario 2-1 were compared to the time series of monthly (or several times per year) measurements taken at A-5 and A-10 in Figure 1. COD, TN and TP, adopted as environmental indexes in Japan, were calculated from the model components as follows.

$$COD = C1 \text{ PhytoPlankton} + C2 \text{ ZooPlankton} + C3 \cdot \text{Detritus} + C4 \text{ DOM} \tag{6.1.1}$$

TN = N_1 · PhytoPlankton + N_2 · ZooPlankton + N_3 · Detritus + N_4 · DOM
+ NH_4-N + $NO_{2,3}$- N   (6.1.2)

TP = P_1 · PhytoPlankton + P_2 · ZooPlankton + P_3 · Detritus + P_4 · DOM +
PO_4-P   (6.1.3)

The mass of Phytoplankton, Zooplankton, Detritus and DOM is expressed as the unit of Carbon mass in the model and the coefficients, C_1 to C_4, N_1 to N_4, and P_1 to P_4 were ratios of COD/C, N/C, and P/C. The values of these coeffcients were given by a multiple regression analysis of the field data. The results of the comparison of the simulated data and the data measured at the observation stations, A-5 and A-10 (see Figure 1) are shown in Figure 12. The observed trends and values are generally well-reproduced by the model. The calculated spatial distributions of the selected pelagic model components at the surface and the cross-section along the dashed line in Figure 1 during summer (averaged value of July, August and September) are shown in Figure 13. The observed vertical stratification of DO in summer is reproduced in the model simulation. The higher concentration at the surface and the lower concentration (hypoxia) at the bottom during summer stand out compared to winter. This trend results from (a) the high rate of photosynthesis (b) the increase of organic matter sedimentation due to the increase in the concentration of phytoplankton and (c) the weak vertical mixing of the water column during the summer.

6.2. EFFECT OF DISAPPEARANCE OF THE NATURAL SEAGRASS BEDS

To clarify the effect of the seagrass beds on the water quality in Atsumi Bay, scenario 2-2 was calculated (no seagrass beds in the Jinno area), and the result was compared to the result of scenario 2-1 (the existing condition). In this calculation, the biomass of seagrasses, epiphytes, and epifauna in the Jinno area was set at zero. The other calculation conditions were the same as the existing conditions, scenario 2-1, described in section 6.1.

The differences in the results of the two scenarios (scenario 2-2 minus scenario 2-1) are shown in Figure 14. This figure shows the differences in June when the biomass of eelgrass is at its maximum.

(a) *Phytoplankton*: The concentration of phytoplankton in the scenario without seagrass beds is higher than the one with seagrass beds in the central area of Atsumi Bay, while it is not changed in the Jinno area. When seagrass beds are present, photosynthesis by eelgrass and epiphytes results in the consumption of the nutrients (during photosynthesis the dissolved inorganic nitrogen and phosphorus is assimilated as organic matter in the plant cell). As a result, nutrient concentrations (NH_4-N, $NO_{3,2}$-N, PO_4-P) in the Jinno area are reduced. On the other hand, in the scenario without seagrass beds (scenario 2-2), the nutrients not utilized by eelgrass and epiphytes in the Jinno area are transported to the central bay area. Then, they are used in the growth of phytoplankton. The reason for no differences in the phytoplankton between scenarios 2-1 and 2-2 in the Jinno area is that nutrients do not become a limiting factor of growth for phytoplankton in the Jinno area. In contrast, the reason for differences in the central bay area is because insufficient nutrients limit the growth of phytoplankton there.

(b) *Detritus*: The concentration of detritus when seagrass beds are present (scenario 2-1) is higher than without seagrass beds (scenario 2-2) around the Jinno area. This result can be attributed to the fluxes in the mortality production by epiphytes and in the mortality and feces production by epifauna associated with seagrass beds. In the central bay area, the higher biomass of phytoplankton in scenario 2-2 leads to its higher mortality. This detritus production is reversed with its reduction due to less transportation of detritus from the Jinno area to the central bay area in scenario 2-2 than scenario 2-1. Therefore, the difference in the concentration of detritus is almost zero.

(c) *NH_4-N*: When seagrass beds are present (scenario 2-1), the concentration of NH_4-N is lower than when they are absent (scenario 2-2). This is due to nutrient fixing during photosynthesis by eelgrass and epiphytes. The spatial distribution and the differences in concentrations of PO_4-P are the same as for NH_4-N. Although increased nutrients of scenario 2-2 in the Jinno area is transported to the central bay area, the differences in the nutrients concentration between scenario 2-1 and 2-2 does not outstand in the central bay due to the utilization for the photosynthesis by increased phytoplankton there in scenario 2-2.

(d) *$NO_{2,3}$-N*: The concentration of $NO_{2,3}$-N is lower when seagrass beds are present. This result is attributed to nutrient fixing during photosynthesis by eelgrass and epiphytes. The nitrification rate in the scenario with seagrass beds (scenario 2-1) is higher compared to the scenario without seagrass beds (scenario 2-2) due to the high oxygen concentrations produced during photosynthesis by eelgrass and epiphytes. However, an increase in the nitrification rate has less effect than a decrease in this rate.

(e) *DO*: The DO concentration in scenario 2-1 (with seagrass beds) is higher than that in scenario 2-2 (without seagrass beds) in the Jinno area. Oxygen is produced during photosynthesis by eelgrass and epiphytes and it is consumed by the many living organisms inhabiting the seagrass beds. However, the rate of oxygen production is higher than the rate of oxygen consumption there. In the central bay area, the DO concentration has no differences between scenarios 2-1 and 2-2, viz: in scenario 2-2, the higher rate of DO production due to the higher concentration of phytoplankton prevents the DO reduction from the Jinno area when compared to scenario 2-1.

6.3. EFFECT OF THE MITIGATION PLAN FAILING IN SEAGRASS GROWING

Scenario 2-3 was calculated to evaluate the effect of the creation of artificial shallows in the Mito area as a mitigation plan for the reclamation of seagrass beds in the Jinno area. In scenario 2-3, seagrass beds at Jinno area is assumed to reclaim, and the creation of artificial shallows is assumed at Mito area. In this scenario, seagrass is assumed not to grow on the artificial shallows in the Mito area. To assess the effect of the mitigation plan failing seagrass growing, scenario 2-3 is compared to the scenario in which the Jinno area is reclaimed and the Mito area is in the existing condition (scenario 2-5). The assumptions of the simulation in scenario 2-3 are as follows:

(a) *Assumption 1*: The depth of the artificial shallows was assumed to be at the datum line minus 0.8 m (D.L. -0.8 m).
(b) *Assumption 2*: The area of the artificial shallows in the Mito area was set as 15 ha and that of the reclamation zone in the Jinno area was set as 37.5 ha.
(c) *Assumption 3*: The natural seagrass beds in the Jinno area were assumed to disappear due to the reclamation.

(d) *Assumption 4*: The initial biomass of benthic fauna (suspension feeders and deposit feeders) was set as the observed value in the Miya area (located 10 km west from the Mito area – refer to Figure. 1). Artificial shallows with seagrass beds growing on them already exist in the Miya area (refer to section 5.3). The biomass of benthic fauna in the Miya area is higher than that in the seagrass beds of the Jinno area.

(e) *Assumption 5*: The initial concentrations of the detritus in sediments and the nutrients in pore waters in the artificial shallows were set at the observed value in the Mito area. The geographical changes due to the reclamation at the Jinno area and the creation of the artificial shallows at the Mito area were set both in the hydrodynamic and ecosystem models.

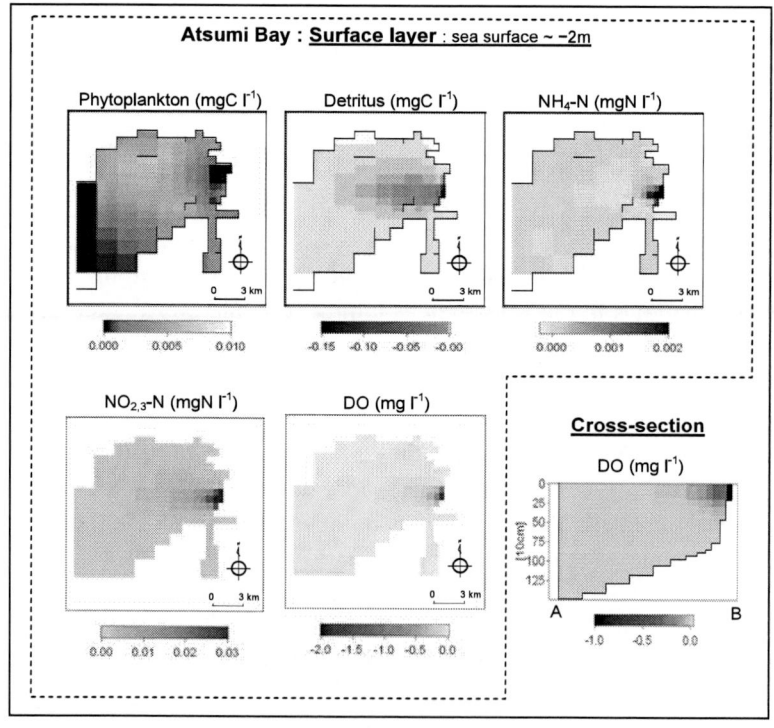

Figure 14. The differences in the concentrations of model components between the scenarios without seagrass and with seagrass in June (scenario 2-2 - scenario 2-1 in Table 2). The location of the cross-section of DO is indicated by a dashed line A-B in Figure 1. The right side of the figure in the cross-section corresponds to point B.

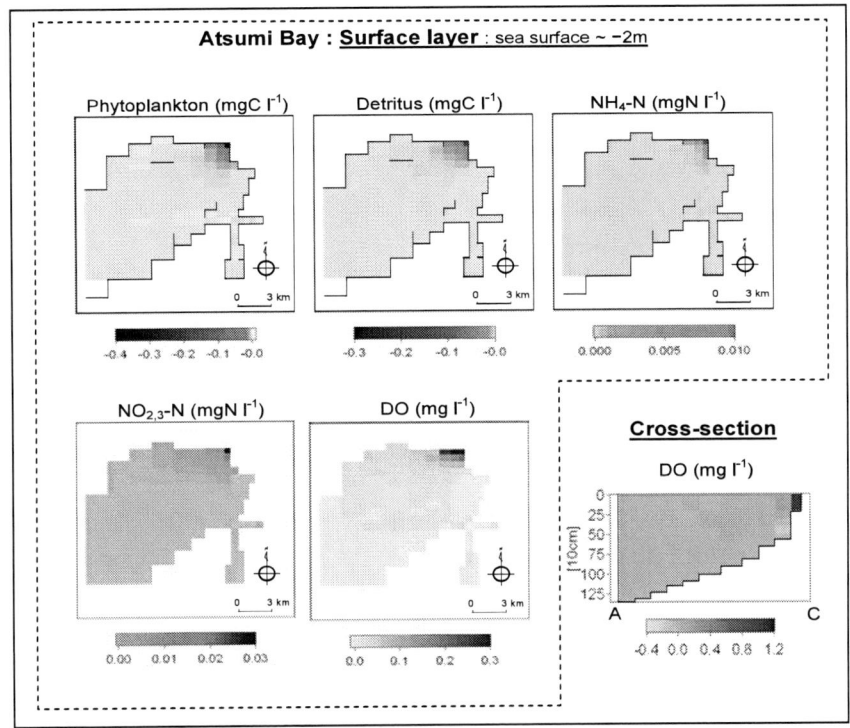

Figure 15. The differences in the concentrations of model components between the presence of artificial shallows and the absence of artificial shallows in summer ("scenario 2-3" minus "scenario 2-5" in Table 2). The location of the cross-section of DO is indicated by a dashed line A-C in Figure. 1. The right side of the figure in the cross-section corresponds to point C.

The other calculation conditions were the same as those for scenario 2-1 (the existing conditions in Atsumi Bay).

The differences in the values of several of the model components between the scenarios with artificial shallows in the Mito area (scenario 2-3) and without artificial shallows in the Mito area (scenario 2-5), i.e. scenario 2-3 minus scenario 2-5, are shown in Figure 15. Figure 15 illustrates the differences during summer (averaged value of July, August, and September). The important role of biological processes in shallow waters without seagrass beds is particularly evident in summer when red tides and hypoxia occur.

(a) *Phytoplankton*: The concentration of phytoplankton in the scenario with artificial shallows is lower than the scenario without artificial shallows. This result is attributed to the increase in suspension feeders (of phytoplankton) in the Mito area due to the creation of the artificial shallows. Excretion by the suspension feeders supplies more nutrients to the water column and has the potential to increase the concentration of phytoplankton if nutrient concentration is a limiting factor of photosynthesis. However, the nutrient concentrations in Atsumi Bay are sufficient for photosynthesis so are not a limiting factor.

(b) *Detritus*: The concentration of detritus in the scenario with artificial shallows is lower than in the scenario without artificial shallows. This result can be largely explained by the increase in the feeding rate of suspension feeders in the artificial shallows. An increase in the concentration of feces from the suspension feeders can contribute to an increase in the concentration of detritus. However, in this scenario the consumption of detritus (feeding) is greater than the production (excretion).

(c) NH_4-N: The concentration of NH_4-N is higher in the scenario with artificial shallows. This result is attributed to the increase in the excretion rate by a greater biomass of suspension feeders and the lower utilization rate of nutrients by photosynthesis of phytoplankton. The biomass of phytoplankton decreases as it is consumed by suspension feeders. In contrast, a decrease cause in the concentration of NH_4-N in the artificial shallows is an increase in its utilization rate due to photosynthesis by benthic algae. The concentration of benthic algae increases with the increase in light intensity at the seafloor compared to without artificial shallows However, the rates of increase exceed the rates of decrease and the net concentration of NH_4-N is higher. The spatial distribution and changes in the concentration of PO_4-P are due to the same factors as NH_4-N.

(d) $NO_{2,3}$-N: The concentration of $NO_{2,3}$-N in the scenario with artificial shallows (scenario 2-3) is higher than without artificial shallows (scenario 2-5). This can be attributed to the increase in the nitrification rate caused by an increase in oxygen and NH_4-N concentrations.

(e) *DO*: The oxygen concentration in the scenario with artificial shallows is higher than without artificial shallows. This is due to the increase in oxygen production by photosynthesis of the benthic algae and the increase in the rate of the vertical mixing of the water column. These increases result from the change in geography (the ground level change to D.L. -0.8 m). The concentration of oxygen can decrease when: (a) there is an increase in the rate of oxygen consumption by the respiration of the greater numbers of suspension feeders in the Mito area; and (b) the rate of oxygen production is reduced by a reduction in the concentration of phytoplankton. In this scenario the rate of oxygen production is greater than the rate of oxygen consumption.

6.4. EFFECT OF THE MITIGATION PLAN SUCCEEDING IN SEAGRASS GROWING

Scenario 2-4 was calculated to evaluate the effect of the creation of artificial shallows in the Mito area as a mitigation plan for the reclamation of the seagrass beds in the Jinno area. In scenario 2-4, seagrass beds at Jinno area is assumed to reclaim, and the artificial shallows is assumed to be created at Mito area. In this scenario, seagrass is assumed to grow on the artificial shallows at Mito area.

To assess the effect of successfully growing seagrass on the artificial shallows, this scenario is compared to scenarios 2-5 in Table 2. The assumptions in the calculation of scenario 2-4 are as follows:

(a) *Assumption 1*: The depth of the artificial shallows is set as D.L. -0.8 m (refer to section 5.3).
(b) *Assumption 2*: The area of the artificial shallows in the Mito area is set as 15 ha and that of the reclamation zone in the Jinno area is set as 37.5 ha.
(c) *Assumption 3*: The natural seagrass beds in the Jinno area are assumed to disappear as a result of the reclamation.
(d) *Assumption 4*: The initial biomass of the benthic fauna (suspension feeders and deposit feeders), and the biomass of seagrasses, epiphytes and epifauna were set as the observed value in the Miya area (located 10 km west from the Mito area – refer to Figure 1), where seagrass beds are already growing on existing artificial shallows (refer to section 5.3).
(e) *Assumption 5*: The initial concentrations of the detritus in sediments and the nutrients in pore waters of the artificial shallows were set at the observed value in the Mito area.

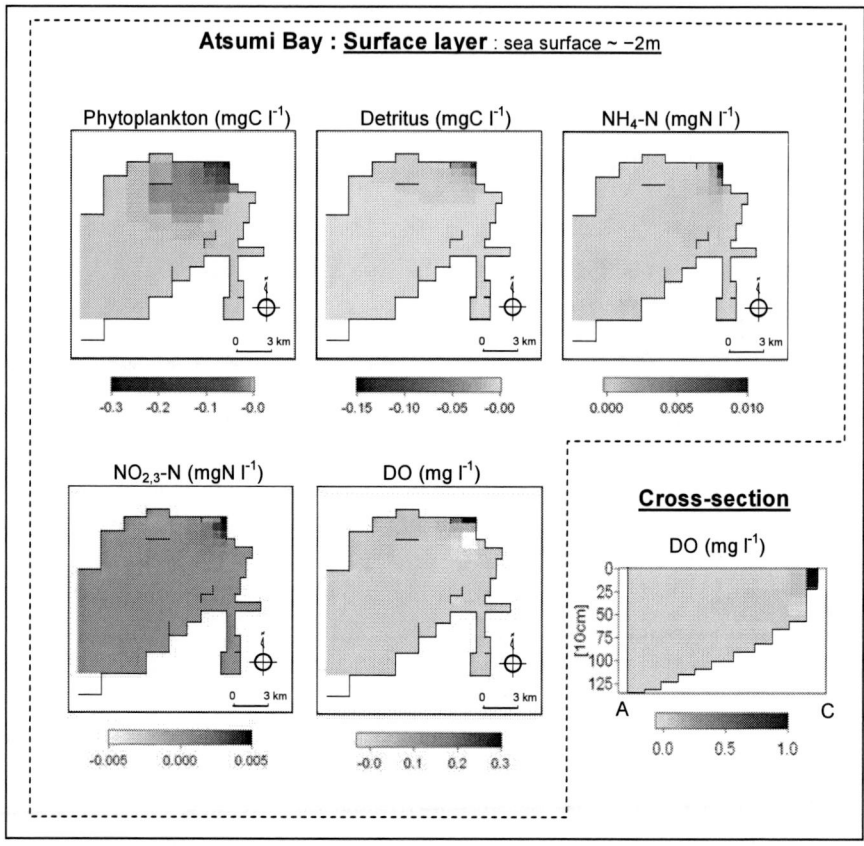

Figure 16. The differences in the concentrations of model components between the presence of artificial shallows with seagrass beds and the absence of artificial shallows in June ("scenario 2-4" minus "scenario 2-5" in Table 2). The location of the cross-section of DO is indicated by a dashed line A-C in Figure 1. The right side of the figure in the cross-section corresponds to point C.

The other calculation conditions were the same as for scenario 2-1 (existing conditions in Atsumi Bay.)

The differences in the values for several model components between scenarios 2-4 and 2-5 for June (biomass of seagrasses is highest during this period) are shown in Figure 16 respectively.

(a) *Phytoplankton*: The concentration of phytoplankton in the scenario with seagrass beds on the artificial shallows at Mito area (scenario 2-4) is lower than in scenario 2-5 (existing condition in the Mito area).

This result is attributed to the same causes as the differences in phytoplankton between scenarios 2-3 and 2-5 (refer to section 6.3).

(b) *Detritus*: The concentration of detritus in the scenario with seagrass beds on artificial shallows (scenario 2-4) is lower than in the scenario without artificial shallows (scenario 2-5). This result can be explained by an increase in the feeding rate of suspension feeders in scenario 2-4. The increase in feces production by the suspension feeders and epifauna in scenario 2-4 could cause an increase in the concentration of detritus. However, the consumption rate of detritus is greater than the production rate and leads to a net increase in the concentration of detritus.

(c) *NH_4-N*: The concentration of NH_4-N is high in the artificial shallows with seagrass beds in the Mito area (scenario 2-4). This increase in concentration is attributed to the increase in the excretion rate of an increasing biomass of suspension feeders. In addition, a decrease in the biomass of phytoplankton results in a lower rate of photosynthesis and utilization of nutrients. Decreases in the concentration of NH_4-N by the creation of artificial shallows with seagrass beds can result from an increase in the rate of nutrient utilization during photosynthesis by seagrasses, epiphytes and benthic algae at Mito area. However, the production rate of NH_4-N is greater than the consumption rate leading to the observed net increase in the concentration of NH_4-N. The spatial distribution and the observed differences in the concentration of PO_4-P are the same as for NH_4-N.

(d) *$NO_{2,3}$-N*: The concentration of $NO_{2,3}$-N in the scenario with seagrass beds on artificial shallows is higher than in the scenario without artificial shallows. This result is attributed to an increase in the nitrification rate caused by the increase in the oxygen and NH_4-N concentrations. The concentration of $NO_{2,3}$-N can decrease as it is utilized during the photosynthesis of seagrasses, epiphytes and benthic algae. However, the rate of utilization is lower than the rate of production.

(e) *DO*: The oxygen concentration in the scenario with seagrass beds on artificial shallows in the Mito area is higher than in the scenario without the artificial shallows. This result is due to the increase in oxygen production during photosynthesis of the benthic algae, additional oxygen produced process (photosynthesis) due to existence of eelgrass and epiphytes, and an increase in the rate of vertical mixing of the water column. These increases result from the change in depth. In addition, the oxygen production by photosynthesis of seagrass and epiphytes are given in the artificial shallows in the Mito area. The creation of the artificial shallows in the Mito area also leads to (a) higher oxygen consumption rates due to the respiration of an increased number of suspension feeders and (b) decrease in the photosynthetic rate of phytoplankton as their numbers are reduced compared to the scenario without the creation of shallows. Both (a) and (b) have the potential to decrease the oxygen concentration. However, the rate of increase is greater than the rate of decrease resulting in a net increase in oxygen concentration.

Chapter 7

EVALUATION OF THE PURIFICATION

7.1. DEFINITION OF THE PURIFICATION RATE

In order to quantitatively evaluate the creation of artificial shallows in the Mito area as a mitigation plan for the reclamation of seagrass beds in the Jinno area, an index is required. The purification rate was defined as this index and analyzed from the results of the ecosystem model calculations shown in sections 5 and 6. The purification rate is the removal flux of each pelagic model components from the estuary, Atsumi Bay. In this section, a comparison is made between the loss of the purification rate through the reclamation of the natural seagrass beds in the Jinno area and the gain in the purification rate through the creation of the artificial shallows at Mito area. The purification rate is calculated from the input fluxes and the output fluxes at the boundary of Jinno and Mito through the advection and diffusion due to the water current. The input fluxes refer to transport from the outside of the Jinno and Mito areas, to the inside of each area, and the output fluxes means the transport from the inside of each area to the outside of each area. The loss of purification rate at Jinno is calculated from the purification rate of the scenario for the reclamation of natural seagrass beds in the Jinno area minus the purification rate of the existing condition in the Jinno area. The gain in the purification rate at Mito is calculated from the purification rate of the scenario for the creation of artificial shallows in the Mito area minus the purification rate of the existing condition in the Mito area.

Table 19. Purification rate evaluated from advection and diffusion (Summer, June, Annual mean value)

Purification component	Jinno area "*Loss*" of purification rate due to the reclamation of natural seagrass beds			Mito area "*Gain*" of purification rate due to the creation of artificial shallows (D.L. -1.5 m)			Mito area "*Gain*" of purification rate due to the creation of artificial shallows without seagrass, epiphytes and epifauna (D.L. -0.8 m)			Mito area "*Gain*" of purification rate due to the creation of artificial shallows with seagrass, epiphytes and epifauna (D.L. -0.8 m)		
Summer/June/ Annual mean value	Sum.	Jun.	Ann. mean	Sum.	Jun.	Ann. mean	Sum.	Jun.	Ann. mean	Sum.	Jun.	Ann. mean
Phyto-plankton (mgC m^{-2} day^{-1})	36.5 (100)	152.8 (100)	268.9 (100)	1168.7 (3201)	1145.8 (749)	1311.0 (487)	2931.0 (8028)	2140.6 (1401)	2706.9 (1007)	2932.3 (8031)	2154.3 (1410)	2707.7 (1007)
Detritus (mgC m^{-2} day^{-1})	110.3 (100)	11.7 (100)	359.2 (100)	608.5 (552)	553.9 (4718)	480.7 (134)	945.6 (857)	964.9 (8217)	832.8 (232)	931.3 (844)	884.0 (7529)	791.5 (220)
TOC (mgC m^{-2} day^{-1})	1070.7 (100)	846.7 (100)	1020.5 (100)	3438.9 (321)	2485.8 (294)	2553.7 (250)	8864.4 (828)	7065.8 (834)	6469.4 (634)	8840.6 (826)	6938.4 (819)	6402.9 (627)
DIN (mgN m^{-2} day^{-1})	-177.1 (100)	206.7 (100)	-133.8 (100)	-206.9 (117)	-96.3 (-47)	-138.8 (104)	-619.6 (350)	-566.8 (-274)	-487.3 (364)	-614.4 (347)	32.6 (16)	-378.5 (283)
DIP (mgP m^{-2} day^{-1})	-26.2 (100)	28.3 (100)	-19.6 (100)	-27.8 (106)	-14.2 (-50)	-18.8 (96)	-81.2 (310)	-77.3 (-273)	-64.4 (328)	-80.5 (308)	-0.5 (-2)	-50.8 (259)

Relative value to loss of purification rate in the Jinno area (Jinno area = 100) is denoted in parentheses. TOC: Total Organic Carbon, DIN: Dissolved Inorganic Nitrogen, DIP: Dissolved Inorganic Phosphorus.

Table 20. Biological purification fluxes (Summer, June, Annual mean value)

Purification fluxes	Jinno area "*Loss*" of biological purification flux due to the reclamation of natural seagrass beds			Mito area "*Gain*" of biological purification flux due to the creation of artificial shallows (D.L. -1.5 m)			Mito area "*Gain*" of biological purification flux due to the creation of artificial shallows without seagrass, epiphytes and epifauna (D.L. -0.8 m)			Mito area "*Gain*" of biological purification flux due to the creation of artificial shallows with seagrass, epiphytes and epifauna (D.L. -0.8 m)		
Summer/June/ Annual mean value	Sum.	Jun.	Ann. mean	Sum	Jun.	Ann. mean	Sum.	Jun.	Ann. mean	Sum.	Jun.	Ann. mean
Biodeposition of POM from the pelagic to the benthic ecosystem due to feeding by benthic fauna (mgC m^{-2} day^{-1})	415	416	580	1981	1792	2025	4207	3176	3865	4290	3550	3999
Phytoplankton (mgC m^{-2} day^{-1})	325	241	399	1360	1057	1382	2886	1806	2644	2885	1757	2636
Detritus in pelagic system (mgC m^{-2} day^{-1})	25	153	147	525	676	582	1238	1295	1147	1324	1718	1288
DIN consumption by benthic algae (mgN m^{-2} day^{-1})	285	183	130	139	67	57	333	116	139	332	115	139
DIP consumption by benthic algae (mgP m^{-2} day^{-1})	38.9	25.0	17.8	17.8	8.6	7.2	42.5	14.8	17.7	42.4	14.7	17.7
DIN consumption by living organisms in seagrass beds (mgN m^{-2} day^{-1})	142	659	153	-	-	-	-	-	-	19	639	123

Table 20. (Continued).

Purification fluxes	Jinno area "Loss" of biological purification flux due to the reclamation of natural seagrass beds			Mito area "Gain" of biological purification flux due to the creation of artificial shallows (D.L. -1.5 m)			Mito area "Gain" of biological purification flux due to the creation of artificial shallows without seagrass, epiphytes and epifauna (D.L. -0.8 m)			Mito area "Gain" of biological purification flux due to the creation of artificial shallows with seagrass, epiphytes and epifauna (D.L. -0.8 m)		
DIP consumption by living organisms in seagrass beds (mgP m^{-2} day^{-1})	19.2	90.5	20.9	-	-	-	-	-	-	2.2	82.4	15.4
De-nitrification (mgN m^{-2} day^{-1})	-5.3	-2.3	-4.2	0.7	0.5	1.0	0.6	1.3	1.7	0.6	0.6	1.6
DIN flux from sediment to water (mgN m^{-2} day^{-1})	-140	-74	-137	70	6	58	113	109	139	116	87	133
DIP flux from sediment to water (mgP m^{-2} day^{-1})	-20.7	-10.7	-19.6	10.9	2.5	8.8	17.5	18.5	20.3	17.9	17.1	19.9

POM: Particulate Organic Matter, DIN: Dissolved Inorganic Nitrogen, DIP: Dissolved Inorganic Phosphorus.

The results of the evaluations of the gain in the purification rate through the creation of artificial shallows in the Mito area and the loss of purification rate through the reclamation of seagrass beds in the Jinno area are shown in Table 19. The calculation scenarios in the Jinno and Mito areas are the same as for Part 2 discussed in section 6 (refer to Table 2) with an additional scenario. The additional scenario shown in Table 19 is the creation of artificial shallows where the depth of the shallows is set at the datum line minus 1.5 m (D.L. -1.5 m). The value of D.L. -1.5 m was based on the observed value in the Mito area that resulted in non-hypoxic conditions throughout the year (Imao et al., 2001. Under this condition, light intensity is insufficient for seagrasses to photosynthesize (less than 3E m^{-2} day^{-1} of the averaged light condition throughout the year.). Therefore, in this scenario the biomass of seagrasses is set at zero.

7.2. BIOLOGICAL PURIFICATION FLUXES

In order to clarify the contribution of the biological processes to the purification rate shown in Table 19, the purification flux due to biological processes was analyzed simultaneously. The selected biological processes in the analysis are the processes that remove the nitrogen and phosphorus from the pelagic ecosystem in the estuary. This means that the selected biological processes are the purification fluxes in terms of eutrophication. The calculated biological purification fluxes are shown in Table 20. The details for the biodeposition of particulate organic matter (POM) by benthic fauna (suspension feeders), the consumption of nutrients by benthic algae and the consumption of nutrients by seagrass beds in Table 20 are as follows.

(a) Biodeposition of POM from the pelagic to the benthic ecosystem due to benthic fauna feeding is the total biodeposition from phytoplankton, zooplankton, and detritus.
(b) The consumption of dissolved inorganic nitrogen and phosphorus (DIN and DIP) by benthic algae is the net consumption i.e. consumption by photosynthesis of benthic algae minus production by benthic algae respiration.
(c) The consumption of DIN and DIP by living organisms in seagrass beds is the net consumption due to the metabolism of eelgrass and epiphytes i.e. consumption by photosynthesis of eelgrass and

epiphytes minus production by excretion due to the respiration of eelgrass and epiphytes.

7.3. PURIFICATION RATE OF PHYTOPLANKTON

The purification rate of phytoplankton, which is directly associated with red tides, is higher than the purification rate of detritus in any scenario of the creation of artificial shallows in the Mito area (Table 19). This result is attributed to the higher concentration of phytoplankton compared to that of detritus in the Mito area. The main process of purification by phytoplankton is the feeding of suspension feeders in the benthic ecosystem (refer to biodeposition of particulate organic matter from the pelagic to the benthic ecosystem due to feeding by benthic fauna in Table 20). The biomass of suspension feeders is affected by the depth of the artificial shallows. The results from the model suggest that a larger biomass of suspension feeders could be supported when the depth of the shallows is higher (D.L. -0.8 m) because of non-hypoxic conditions at the seafloor. Oxygen concentration at the seafloor of artificial shallows is higher when the depth of the shallows is higher. The association between depth and oxygen concentration results from the following. First, the production of oxygen by the photosynthesis of benthic algae increases even if seagrass beds are not present. Second, the vertical mixing process in the scenario with the creation of artificial shallows with a D.L. -0.8 m is stronger than in the scenario with a D.L. -1.5 m and/or the existing condition in the Mito area.

7.4. PURIFICATION RATE OF NUTRIENTS

The values for the loss of purification rate of nutrients due to the reclamation in the Jinno area (Table 19) are negative both in summer and for the annual average. This result means that more nutrients are removed from the pelagic ecosystem of Atsumi Bay in the scenario of the reclamation in the Jinno area compared to the scenario of the existing conditions in the Jinno area. This result is due to the disappearance of nutrient fluxes from the benthic to pelagic ecosystems, i.e. the reclamation makes the sediment-water interface disappear (Table 20). In contrast, in June, the loss of purification rate of nutrients due to the reclamation in the Jinno area (Table 19) is a positive value.

That is, in June, fewer nutrients are removed from the pelagic ecosystem of Atsumi Bay in the scenario of the reclamation in the Jinno area compared to the scenario of the existing conditions for the Jinno area. This leads to a reduction in the water quality and eutrophication. This result is due to the reduction in nutrient consumption by photosynthesis of eelgrass and epiphytes because of the reclamation of the seagrass beds in the Jinno area (Table 20). Nutrient consumption in seagrass beds due to photosynthesis by eelgrass and epiphytes in the Jinno area is highest in June when the biomass of eelgrass and epiphytes is also highest.

The values for the gain of purification rate due to the creation of artificial shallows in the Mito area (Table 19), are both positive (an improvement in water quality) and negative (a decline in water quality). The positive value for the gain of purification rate is due to nutrient utilization during photosynthesis of seagrasses and epiphytes (Table 20). The negative value is due to an increase of nutrient flux as a result of an increase in the transportation rate of organic matter from the pelagic to the benthic ecosystem by suspension feeders. That is, organic matter is accumulated in the benthic ecosystem of the artificial shallows in the Mito area instead of being transported outside this area by the action of water currents in the pelagic ecosystem.

From the comparison between the scenario of artificial shallows with seagrass beds in the Mito area and the scenario of artificial shallows without seagrass beds in the Mito area (Table 19), the purification rates of the nutrients in the scenario with seagrass beds are higher than without seagrass beds. This result is attributed to the effect of nutrient utilization during photosynthesis of eelgrass and epiphytes (Table 20).

Chapter 8

CONCLUSION

This document describes the concept and numerical construction of an ecosysmtem model and its implementation for the environmental assessment of seagrass beds. The model is the benthic and pelagic coupled ecosystem model and can demonstrate the dynamics of coastal ecosystems such as hypoxic/seagrass/tidal flat ecosystem in terms of carbon, nitrogen, phosphorus and oxygen cycles.

The ecosystem model was used to evaluate (a) the likelihood of the growth of seagrass beds on artificial shallows in the Mito area and (b) the effects of the reclamation of seagrass beds in the Jinno area and the creation of artificial shallows in Mito area on the water quality of the estuary in Atsumi Bay. In order to evaluate the likelihood of the growth of seagrass beds in artificial shallows, first, the characteristics of the nutrient cycling mechanisms in natural seagrass beds were investigated by comparing three scenarios for the Jinno area: sea grass beds, shallow waters without seagrass beds and a hypoxic area. The results of this analysis suggest that biological processes are the major driving forces of nutrient cycling in seagrass beds compared to the other two areas. Second, the effect of the biological processes of the benthic fauna (suspension feeders) on photosynthetic rate (specific growth rate) of seagrass was evaluated through the comparison of two scenarios: (a) seagrass beds wthout suspension feeders in the Jinno area; and (b) the existing condition (seagrass beds with suspension feeders) in the Jinno area. This comparison revealed that suspension feeders made the deeper euphotic zone and led to the higher concentration of the nutrients. As a result, the rate of photosynthesis in seagrass beds with suspension feeders is higher than in those without suspension feeders. Third, the growth of seagrass beds on artificial shallows in the Mito area was evaluated. This analysis suggested that the likelihood of seagrass growth on the artificial shallows in Mito area is high if the depth of the shallows was set at D.L. -0.8 m.

In order to evaluate the effect of the reclamation in the Jinno area and the creation of artificial shallows in the Mito area on the water quality of the estuary, several scenarios were calculated by applying the ecosystem model to Atsumi Bay (including the Jinno and Mito areas). The scenarios calculated by the model were (a) with/without seagrass beds in the Jinno area, (b) the reclamation in the Jinno area and (c) the creation of artificial shallows in the Mito area. The model could describe the differences of the dynamics and spatial distribution of the water quality in Atsumi Bay for each scenario. Finally, the purification rate was defined to quantitatively evaluate the mitigation effect of the creation of artificial shallows in the Mito area on the reclamation of the seagrass beds in the Jinno area. The loss and gain of the purification rate were analyzed for two scenarios: (a) the reclamation in the Jinno area; and (b) the creation of artificial shallows in the Mito area.

Although there are several assumptions underlying these simulation results, the evaluations demonstrated in this paper using the ecosystem model can be used as a reference or base for any discussions or decisions regarding reclamation or mitigation plans for the coastal zone.

Developing this new model means having a good tool to estimate (predict) the effect or the possibility of coastal environmental management, including the re-creation of artificial sea-grass beds.

ACKNOWLEDGMENT

Part of this research was supported by the Program for Promoting Fundamental Transport Technology Research from the Japan Railway Construction, Transport and Technology Agency (JRTT).

REFERENCES

Admiraal, W., Peletier, H., Zomer, H., 1982. Observations and experiments on the population dynamics of epipelic diatoms from an esturaine mudflat. Estuarine, *Coastal and Shelf Science*, 14, 471-487.

Aoyama, H, Suzuki, T, 1997. In Situ Measurement of Particulate Organic Matter Removal Reates by a Tidal Flat Macrobenthic Community. *Bulletin of the Japanese Society of Fisheries Oceanography*, 61(3), 265-274. (in Japanese with English abstract)

Baretta, J.W., Ruardij, P., 1988. Tidal flat estuaries, simulation and analysis of the Ems Estuary. *Ecological Studies*, 71, Springer-Verlag.

Baretta, J.W., Ebenhöh, W., Ruardij, P., 1995. The European Regional Seas Ecosystem Model, a complex marine ecosystem model. *Neth. J. Sea Res.* 33, 233-246.

Baretta-Bekker, J.G., Baretta, J.W., 1997. Microbial dynamics in the marine ecosystem model ERSEM II with decoupled carbon assimilation and nutrient uptake. *J. Sea Res.*, 38, 195-211.

Berg, P., Risgaard-Petersen, N. and Rysgaard, S., 1998. Interpretation of measured concentration profiles in sediment pore water. *Limnology and Oceanography*, 43, 1500-1510.

Berg, P., Røy, H., Janssen F., Meyer. V., Jørgensen, B.B., Huettel, M. and Beer D., 2003. Oxygen uptake by aquatic sediments measured with a novel non-invasive eddy-correlation technique. *Marine Ecology Progress Series*, 261, 75-83.

Berner, R.A., 1980. *Early diagenesis - A theoretical approach -*. Princeton University Press, New Jersey, 241pp.

Boudreau, B.P., 1996. A method of lines code for carbon and nutrient diagenesis in aquatic sediments. *Computers and Geosciences* 22(5), 479-496.

Boudreau, B.P. and Jørgensen, B.B.(eds.) 2001. The Benthic Boundary Layer. *Transport Processes and Biogeochemistry.* Oxford University Press, Oxford, 440pp.

Blumberg. A.F., G.L. Mellor, 1978. A coastal ocean numerical model. In: J. Sundermann and K.P. Holz (Editors), *Mathematical Modeling of Estuarine Physics, Proceedings of an International Symposium*, Hamburg, August 24 to 26, 1978, Springer-Verlag, Berlin, 203-219.

Cammen, L.M., 1980. Ingestion rate: an empirical model for aquatic deposit feeders and detritivores. *Oecologia*, 44, 303-310.

Canfield, D.E., Jørgensen, B.B., Fossing, H., Glud, R., Gundersen, J., Ramsing, N.B., Thamdrup, B., Hansen, J.W., Nielsen, L.P. and Hall, P.O.J., 1993. Pathways of organic carbon oxidation in three continental margin sediments, *Marine Geology,* 113, 27-40.

Chiba Prefecture, 1998-2002. *Result of the water quality survey of public water areas.* (in Japanese)

Chiba Prefectural Fisheries Research Center, 2001-2006. *Hypoxia quick information.* (web site http://www.awa.or.jp/home/cbsuishi/04tkhinsanso/04tkhinsansoflame.htm) (in Japanese)

Conover, R.J., 1978. Transformation of organic matter. In: Kinne, O. (Editor), *Marine Ecology,* vol. IV. Dynamics, Wiley, New York, pp.221-499.

de Beer, D., Wenzhöfer, F., Ferdelman, T.G., Boehme, S.E., Huettel1, M., van Beusekom, J.E.E., Böttcher, M.E., Musat, N., Dubilier, N., 2005. Transport and mineralization rates in North Sea sandy intertidal sediments, Sylt-Rømø Basin, Wadden Sea. *Limnology and Oceanography,* 50, 113-127.

Dedieu, K., Rabouille, G., Gilbert, F., Soetaert, K., Metzger, E., Simonucci, G., Jézéquel, D., Prévot, F., Anschutz, P., Hulth, S.,Ogier, S., Mesnage, V., 2007. Coupling of carbon, nigrogen and oxygen cycles in sediments from a Mediterranian lagoon: a seasonal prespective. *Marine Ecology Progress Series* 346, 45-59.

Emerson, S. and Hedges, J.L., 1988. Processes controlling the organic carbon content of open ocean sediments. *Paleoceanography,* 3, 621-634.

Epply, R.W., Rogers, J.N., McCarthy, J.J., 1969. Half saturation constants for uptake of nitrate and ammonium by marine phytoplankton. *Limnology and Oceanography* 14, 912–920.

Fuhs, W.G., Demmerle, S.D., Canelli, E. and Chen, M., 1972. Characterization of phosphorus-limited plankton algae (with reflections on the limiting nutrient concept). In: Likens, G. E. (Editor), Nutrients and Eutrophication. *Spec. Symp. Vol. 1. Am. Soc. Limnology and Oceanography*, pp.113-133. Allen Press, Lawrence, KS.

Furota, T., 1988. Effects of low-oxygen water on benthic and sessile animal communities in Tokyo Bay. In Symposium: Material cycling and biological environment in Tokyo Bay. *Bulletin on Coastal Oceanography.* 25(2) 104-113. (in Japanese)

Gundersen, J.K., Glud, R.N., Jørgensen, B.B., 1995. Oxygen transformations in the sea floor (in Danish). Marine Research from the Danish Environmental Agency, Vol. 57.

Hata K. and Nakata, K., 1998. Evaluation of eelgrass bed nitrogen cycle using an ecosystem model. *Environmental Modeling & Software.* 13, 491-502.

Hiroshima Environment & Health Association, 2002. *Report about ecological improvement of sediment quality using benthic algae in Seto Inland Sea* (web site http://nippon.zaidan.info/seikabutsu/2001/00614/mokuji.htm) (in Japanese)

Hiwatari, T., Kohata, K. and Iijima, A., 2002. Nitrogen budget of the bivalve Mactra veneriformis, and its significance in benthic-pelagic systems in the Sanbanse area of Tokyo Bay. *Estuar. Coast. Self Sci.,* 55, 299-308.

Horiguchi, F., 2001. Numerical simulations of seasonal cycle of Tokyo Bay using an ecosystem model. *Journal of Advanced Marine Science and Technology Society,* 7(1&2), 1-30. (in Japanese with English abstract)

Imao, K., Suzuki, T., Takabe, T., 2004. New method to predict changes in the structure and function of a macrobenthic community from changes in environmental oxygen concentrations. *Fisheries Engineering,* 41(1), 13-24. (in Japanese with English abstract)

Ishida, M., Hara, T., 1996. Changes in water quality and eutrophication in Ise and Mikawa Bays. *Bulletin of the Aichi Fisheries Research Institute,* 3, 29-41. (in Japanese with English abstract)

Ishikawa, M. and Nishimura, H., 1983. A new method of evaluating the mineralization of particulate and dissolved photoassimilated organic matter. *Journal of the Oceonography Society of Japan,* 39 (2), 29–42.

Isono, R., Kita, J., Kishida, C., 1998. Upper temperature effect on rates of growth and oxygen consumption of the Japanese little neck clam, Ruditapes philippinarum. *Journal of the Oceanography Society of Japan*, 39(2), 29-42.

Japan Environmental Management Association for Industry, 1998. *Survey report of the water quality pollution mechanism in Mikawa Bay.* (in Japanese)

Jørgensen, B.B., 1978. A comparison of methods for the quantification of bacterial sulfate reduction in coastal marine sediments: II Calculations from mathematical models. *Geomicrobiology Journal* 1, 29-47.

Jørgensen, S.E. (Ed.), 1979. Handbook of Environmental Data and Ecological Parameters. International Society for Ecological Modelling, *Pergamon Press, Amsterdam,* 1162 pp.

Jørgensen, S.E., Nielsen, S.N. and Jørgensen, L.A., 1991. *Handbook of Ecological Parameters and Ecotoxicology.* Elsevier Science Publishers, Amsterdam, 1263 pp.

Jørgensen, S.E. and Bendoricchio, G., 2001. *Fundamentals of ecological modelling: Developments in environmental modelling,* 21, 3rd ed. Elsevier, New York, 530pp.

Kakino J., 1982. *Effects of Ao-Shio on the mortality in Manila clams.* Bulletin of Chiba Prefectural Fisheries Research Institute, 40, 1-6. (in Japanese)

Kamio, K., Nomura, M., Nakamura, Y., Kuwae, T., Inoue, T., Konuma, S., 2004. Oxygen variation of the tidal flat overlying water. In: *Proceedings of the 2004 Spring Annual Meeting of the Oceanographic Society of Japan,* p199. (in Japanese)

Kanagawa Prefecture, 1998-2002. *Result of the water quality survey of public water areas.* (in Japanese)

Kanagawa Prefectural Fisheries Research Institute, 2005-2006. *Tokyo Bay Dissolved Oxygen Information.* (web site http://www.agri.pref.kanagawa.jp/suisoken/kankyo/sanso/TokyoBayOxInfo.htm) (in Japanese)

Kikuchi, T., 1993. Ecological characteristics of the tidal flat ecosystem and importance of its conservation. *Japanese Journal of Ecology,* 43, 223-235. (in Japanese)

Koike, K., 2000., *Reclamation of Tokyo Bay and artificial beach in Kanto and Ogasawara areas - Japanese geography -.* University of Tokyo Press. (in Japanese)

Kremer, J.N. and Nixon, S.W., 1978. *A coastal marine ecosystem. simulation and analysis.* Springer-Verlag, Berlin. 217pp.

Kurashige, H., 1942. Resistance of Paphia philippinarum Adams et Reeve to Lack of Oxygen. *Journal of the Oceanographical Society of Japan,* 1 (Nos. 1-2), 123-132. (in Japanese)

Kuwae, T., 2001. Biogeochemical roles of benthic microorganisms in intertidal sandflats. *Ph. D. thesis,* Kyoto University, 93pp.

Kuwae, T., Kibe, E. and Nakamura, Y., 2003. Effect of emersion and immersion on the porewater nutrient dynamics of an intertidal sandflat in Tokyo Bay. *Estuarine, Coastal and Shelf Science,* 57, 929-940.

Kuwae, T., Inoue, T., Miyoshi, E., Konuma, S., Hosokawa, S., Nakamura, Y., 2005. In: Modeling the coastal marine ecosystem coupled with tidal flats based on the study of oxygen cycling in sediments. *Report of Program for Promoting Fundamental Transport Technology Research.* pp.262-423. Japan Railway Construction, Transport and Technology Agency. (in Japanese)

Kuwae, T., Kamio, K., Inoue, T., Miyoshi, E. and Uchiyama, Y., 2006. In situ measurement of oxygen exchange flux between sediment and water of an intertidal sandflat, measured in situ by the eddy-correlation method. *Marine Ecology Progress Series,* 307, 59-68.

Luff, R. and Moll, A., 2004. Seasonal dynamics of the North Sea sediments using a three-dimensional coupled sediment-water model system. *Continental Shelf Research,* 24, 1099-1127.

Marshall, S.M., Orr, A.P., 1955a. Experimental feeding of the copepod Calanus finmarchicus on phytoplankton cultures labeled with radioactive carbon. *Pap. Marine Biology and Oceanography, Deep-Sea Research* 3(Suppl.), 110-114.

Marshall, S.M., Orr, A.P., 1955b, On the biology of Calanus finmarchicus. VIII. Food uptake, assimilation and excretion in adult and stage V. Calanus. *Journal of the Marine Biological Association of United Kindom,* 34, 495-529.

Matsumoto, E., 1983. The sedimentary environment in Tokyo Bay. *Earth Chemistory,* 17, 27-32. (in Japanese)

Matsunaga, K., 1981. Studies on the decomposition processes of phytoplanktonic organic matter. *The Japanese Journal of Limnology,* Vol. 42 (4), 220-229.

Mellor, G.L., Yamada, T., 1982. Development of a turbulent closure model for geophysical fluid problems. *Reviews of Geophysics,* 20, 851-875.

Macedo, M.F., Duarte, P., Mendes, P., Ferreira, J.G., 2001. Annual variation of environmental variables, Phytoplankton species composition and phytosynthetic parameters in a coastal lagoon. *Journal of Plankton Research,* 23(7), 719-732.

Ministry of the Environment, 2006. *Reference data of a basic principle on the regulation of total amount control for chemical oxygen demand, contained amount of nitrogen and phosphorus.* (in Japanese)

Ministry of the Environment, 1998-2002. Comprehensive survey on regional water quality. (in Japanese)

Miyata, M., 2003. Bibliography of Zappai historical sources (*Zappai shiryou kaidai*). Systemized Japanese historical bibliography, *Seisyohdosyoten*, pp.501.

Nakamura, Y., Nomura, M., Kamio, K., 2004. Field observation and analysis of benthic–pelagic coupling in Banzu tidal flat and the adjoincent coastal area of Tokyo Bay. *Report of the Port and Airport Research Institute*, 43(2), 35-71. (in Japanese with English abstract)

Nakata, K., Horiguchi, F., Taguchi, K., Setoguchi, Y., 1983a. Three dimensional simulation of tidal current in Oppa Estuary. *Bulletin of the National Research Institute for Pollution and Resources*, 12(3), 17-36. (in Japanese)

Nakata, K., Horiguchi, F., Taguchi, K., Setoguchi, Y., 1983b. Three dmensional eco-hydrodynamical model in coastal region. *Bulletin of the National Research Institute for Pollution and Resources*, 13(2), 119-134. (in Japanese)

National Institute for Land and Infrastructure Management, 2006. *Integrated Environmental Monitoring at Tokyo Bay* (2002-2003). (web site http://www.nilim.go.jp/) (in Japanese)

Nishikawa, T., Miyahara, K., Nagai S., 2002. The growth response of Coscinodiscus wailesii Gran (Bacillariophyceae) as a function of irradiance isolated from Harima-Nada, Seto Inland Sea, Japan. *Bull. Plankton Soc.* 49(1), 1-8.(in Japanese with English abstract)

Odum, E.P., 1971. *Fundamentals of ecology,* 3rd ed. W.B. Saunders, Philadelphia.

Ogura, N., 1972. Decomposition of dissolved organic matter derived from dead phytoplankton. pp.507-515. In Takenouti, A. Y. (ed.) *Biological Oceanography of the Northern Pacific Ocean.* Idemitsu Shoten, Tokyo.

Ogura, N., 1975. Decomposition of dissolved organic matter in coastal seawater. *Marine Biology* 31, 101–111.

Oguz, T., 2002. The role of physical processes controlling the oxycline and suboxic layer structures in the Black Sea, *Glob. Biogeochem. Cycles,* 16(2), 101029-101042.

Patankar, S.V., 1980. Numerical heat transfer and fluid flow. *Hemisphere Publishing,* USA, pp.1-197.

Revsbech, N. P., Madsen, B. and Jørgensen, B.B., 1986. Oxygen production and consumption in sediments determined at high spatial resolution by computer simulation of oxygen microelectrode data. *Limnology and Oceanography,* 31(2), 293-304.

Rosenfeld, J.K., 1979. Ammonium adsorption in nearshore anoxic sediments. *Limnology and Oceanography,* 24, 356-364.

Rysgaard, S. and Berg, P., 1996. Mineralization in a northeastern Greenland sediment : mathematical modeling, measured sediment pore water profiles and actual activities, *Aquatic Microbial Ecology,* 11, 297-305.

Sayama, M., 2005. In: Modeling the coastal marine ecosystem coupled with tidal flats based on the study of oxygen cycling in sediments. Report of Program for Promoting Fundamental Transport Technology Research. pp.424-456. Japan Railway Construction, *Transport and Technology Agency.* (in Japanese)

Soetaert, K., Herman, P.M.J., Middleburg, J.J., 1996a. A model of early diagenetic processes from the shelf to abyssal depth, *Geochimica et Cosmochimica Acta* 60(6), 1019-1040.

Soetaert, K., Herman, P.M.J., Middleburg, J.J., 1996b. Dynamic response of deep-sea sediments to seasonal variations: *Amodel. Limnol. Oceanogr.,* 41(8), 1651-1668.

Soetaert, K., Middelburg, J.J., Herman, P.M.J., Buis, K., 2000. On the coupling of benthic and pelagic biogeochemical models. *Earth-Science Reviews,* 51, 173-201.

Sohma, A., Sato, T., Nakata, K., 2000. New numerical model study on a tidal flat system - seasonal, daily and tidal variation. *Spill Science Technology Bulletin,* 6, 173-185.

Sohma, A., Sekiguchi, Y., Yamada, H., Sato, T., Nakata, K., 2001. A new coastal marine ecosystem model study coupled with hydrodynamics and tidal flat ecosystem effect, *Marine Pollution Bulletin,* 43, 187-208.

Sohma, A., Sayama, M., 2002. Modeling for coupled cycle of Oxygen, Nitrogen, and Carbon in a coastal marine sediment - A new ecological model for dynamics in the micro profiles -. In: *Proceedings of Coastal Engineering, JSCE,* 49, 1231-1235. (in Japanese)

Sohma, A., Sekiguchi, Y., 2003. Development of a new multiple coastal ecosystem model focused on ecological network and benthic vertical mechanisms in the micro scale - application of a hydrodynamics model and benthic ecosystem model in the central bay area of Tokyo Bay -. *Proceedings of Advanced Marine Science and Technology conference in autumn,* 87-92. (in Japanese)

Sohma, A., Sekiguchi, Y., Nakata, K., 2004. Modeling and evaluating the ecosystem of sea-grass beds, shallow waters without sea-grass, and an oxygen-depleted offshore area. *Journal of Marine Systems*, 45, 105-142.

Sohma, A., Sekiguchi, Y., Kakio, T., 2005a. Development of a new multiple coastal ecosystem Model "ZAPPAI" including benthic, pelagic and tidal flat ecosystems for ecological evaluation in hypoxic estuary. - autonomous response to the tidal flat creation, dredging, sand capping, load reduction and red tide - , *Journal of Advanced Marine Science and Technology Society*, 11, 2, 21-52 (in Japanese with English abstract)

Sohma, A., 2005b. *Development of a multiple coastal ecosystem model including benthic, pelagic and tidal flat ecosystems for ecological evaluation in hypoxic estuary.* Ph. D. thesis, Tokai University, 368pp.

Sohma, A., Sekiguchi, Y., Kuwae, T., Nakamura, Y., 2008. A Benthic-pelagic coupled ecosystem model to estimate the hypoxic estuary including tidal flats - model description and validation of seasonal/daily dynamics -. *Ecological Modeling* 215, 10-39.

Spalding, D.B., 1972. A novel finite-difference formulation for differential expressions involving both first and second derivatives. *International Journal for Numerical Methods in Engineering*, 4, 551-559.

Strickland, J.D.H., 1965. Chemical composition of phytoplankton and method for measuring plant bio-mass, practical considerations composition ratios. *Chemical Oceanography*, 1, 514-518.

Suschenya, L.M., 1970. Food rations, metabolism, and growth of crustaceans. In: Steele, J.H. (Editor), *Marine Food Chains*, University of California Press, Berkeley, CA.

Suzuki, T., Aoyama, H., Kai, M., Imao, K., 1998. Effect of dissolved oxygen deficiency on a shallow benthic community in an embayment. *Oceanography in Japan*, 7(4), 223-236. (in Japanese with English abstract)

Suzumura, M., Kokubun, H., Itoh, M., 2003. Phosphorus cycling at the sediment-water interface in a eutrophic environmnet of Tokyo Bay, Japan. *Oceanography in Japan*, 12(5), 501-516. (in Japanese with English abstract)

Tokyo Metropolitan, 1998-2002. *Result of the water quality survey of public water areas.* (in Japanese)

Valiela, I, 1984. *Marine Ecological Processes,* 1-546pp, Springer, New York.

Yamamuro, M., Koike, I., 1993. Nitrogen metabolism of the filter feeding bivalve Corbicula Japonica and its significance in primary production of a brackish lake in Japan. *Limnol. Oceanogr.,* 38, 997-1007.

Zillioux, E., 1970. Ingestion and assimilation in laboratory cultures of acartia. *Technical Report, the National Marine Water Quality Laboratory*, EPA, Narragansett, RI.

INDEX

A

adsorption, 21, 107
algorithm, 13, 17, 23
ammonium, 40, 102
assimilation, 35, 101, 105, 109

B

benthic algae 3, 14, 16, 23, 38, 39, 42, 52, 53, 56, 59, 62, 64, 65, 69, 84, 87, 91, 93, 94, 103
benthic fauna, vii, 1, 2, 5, 70, 72, 73, 82, 85, 91, 93, 94, 97
biochemical reactions, 14, 17, 21, 23, 27, 44, 45
biological processes, 7, 8, 14, 62, 67, 68, 69, 70, 83, 93, 97
boundary conditions, 16, 43, 44, 57, 77

C

coordination, 17
cycles, vii, 14, 16, 58, 97, 102

D

decomposition, 105
degradation, 30, 37, 47, 51
denitrification, 30, 65
deposition, 1, 14, 16, 70
desorption, 21
dissolved oxygen, 41, 42, 53, 59, 78, 108

E

eelgrass 2, 5, 31, 44, 48, 60, 64, 65, 70, 71, 73, 74, 79, 80, 81, 87, 93, 94, 95, 103
environmental assessment, 3, 97
ecosystem, vii, 2, 3, 5, 7, 9, 10, 11, 13, 14, 15, 16, 17, 19, 43, 44, 57, 71, 73, 82, 89, 91, 93, 94, 95, 97, 98, 101, 103, 104, 105, 107, 108
environmental impact, 5
excretion, 7, 29, 32, 50, 64, 65, 69, 70, 73, 84, 87, 94, 105

F

feces, 1, 7, 28, 29, 32, 65, 71, 80, 84, 87
fish, 64, 65
flow field, 13

G

geography, 84, 104
grids, 58, 72, 77

growth rate, vii, 7, 29, 31, 32, 34, 38, 46, 48, 49, 50, 52, 70, 73, 97

H

half saturation constant, 46, 47, 48, 49, 50, 51, 52
hypoxia, 1, 2, 72, 79, 83

I

immersion, 105
inhibition, 27, 47, 51
intraphase mixing, 20
interphase,mixing 20, 45, 52
iron, 27

L

light conditions, 71

M

management, 2, 98
manganese, 27
metabolism, 50, 70, 93, 108
mortality rate, 50
multiple regression analysis, 79

N

nitrate, 27, 30, 37, 40, 56, 102
nitrification, 30, 42, 47, 51, 54, 56, 59, 81, 84, 87, 92
nitrogen, 2, 14, 15, 16, 40, 46, 48, 49, 52, 58, 59, 62, 64, 66, 67, 68, 69, 78, 80, 93, 97, 103, 105
nitrogen gas, 40
nutrients, 1, 2, 44, 54, 58, 62, 70, 71, 72, 73, 80, 82, 84, 85, 87, 93, 94, 95, 97
nutrient cycling, 58, 68, 69, 97

O

organic matter, 1, 17, 23, 27, 29, 40, 59, 62, 71, 73, 79, 80, 93, 94, 95, 102, 103, 105, 106
organism, 45
oscillation, 53, 56
oxidation, 28, 30, 42, 47, 52, 102
oxygen, 2, 5, 7, 14, 15, 16, 27, 33, 34, 35, 41, 42, 50, 53, 56, 57, 58, 59, 62, 63, 67, 69, 78, 81, 84, 87, 94, 97, 102, 103, 105, 107, 108
oxygen consumption, 41, 81, 84, 88, 103

P

phosphorus, 14, 16, 46, 48, 49, 52, 58, 78, 80, 93, 97, 103, 105
photosynthesis, 1, 7, 14, 27, 29, 31, 42, 48, 52, 53, 56, 59, 62, 65, 69, 70, 71, 73, 79, 80, 81, 84, 87, 93, 94, 95, 97
phytoplankton, viii, 1, 17, 28, 46, 53, 59, 64, 65, 70, 73, 79, 80, 81, 83-88, 93, 94, 102, 105, 106, 108
porosity, 20, 38, 45
purification, vii, 89-95, 98

R

reaction rate, 27
resolution, 43, 77, 106
respiration, 28, 29, 31, 32, 42, 46, 48, 49, 52, 64, 65, 85, 88, 93, 94

S

salinity, 44
seagrass, vii, viii, 1, 2, 5-11, 14, 57, 58, 69-74, 77, 79, 80, 81, 82, 83, 85-95, 97, 98
seagrass cultivation, 9, 11, 72
sediment, 14, 16, 20, 21, 23, 43, 45, 59, 64, 65, 92, 94, 101, 103, 105, 107, 108
sedimentation, 1, 64, 70, 79

sediments, 1, 45, 70, 71, 72, 82, 85, 101, 102, 104, 105, 106, 107
sediment-water interface, 16, 20, 64, 65, 94, 108
shoot, 11
simulation, 13, 43, 44, 72, 73, 77, 79, 81, 98, 101, 104, 106
stratification, 79
surface layer, 41, 44

T

temperature, 13, 16, 27, 31, 34, 35, 44, 45, 48, 50, 56, 103
tides, 1, 83, 94
time series, 43, 53, 78
transparency, vii, 3, 13, 62, 71
transplantation, 73
turbulence, 13
turnover, vii, 59, 62, 67, 68, 69

U

urban area, 2

V

validation, 57, 108
vegetation, 1, 2, 69
velocity, 13, 14, 16, 20, 59
vertical mixing, 69, 79, 84, 87, 94
viscosity, 14, 17

W

water quality, vii, 9, 11, 13, 73, 79, 95, 97, 98, 102, 103, 104, 106, 108

Z

zooplankton, 17, 29, 46, 64, 93